A GEOGRAPHY OF energy

THE BROWN
FOUNDATIONS OF GEOGRAPHY
SERIES

Consulting Editor
ROBERT H. FUSON
University of South Florida

A GEOGRAPHY OF

Agriculture
James R. Anderson, University of Florida

Soils
Robert M. Basile, University of Toledo

Transportation and Business Logistics
J. Edwin Becht, The University of Oklahoma

Plants and Animals
David J. de Laubenfels, Syracuse University

Geography
Robert H. Fuson, University of South Florida

The Atmosphere—(Second Edition)
John J. Hidore, Indiana University

Politics
W. A. Douglas Jackson, The University of Washington
Edward F. Bergman, Herbert H. Lehman College

Regions
James R. McDonald, Eastern Michigan University

Population and Settlement
Maurice E. McGaugh, Central Michigan University

Industrial Location
E. Willard Miller, The Pennsylvania State University

Water
Ralph E. Olson, The University of Oklahoma

Earth Form—(Second Edition)
Stuart C. Rothwell, University of South Florida

Landforms
John H. Vann, California State College, Hayward

Enery
H. Reid Wagstaff, Arizona State University

THE BROWN
FOUNDATIONS OF GEOGRAPHY
SERIES

A GEOGRAPHY OF
energy

H. REID WAGSTAFF
Arizona State University

WM. C. BROWN COMPANY PUBLISHERS
DUBUQUE, IOWA

FOUNDATIONS OF GEOGRAPHY

Consulting Editor

Robert H. Fuson
University of South Florida

Library of Congress Catalog Card Number: 73—88488

ISBN 0—697—05168—4

Printed in the United States of America

Geography is one of man's oldest sciences, yet it is as new as the Space Age. Knowledge of the earth obtained from satellite photography and measurement, remote sensing of the environment, and by means of other sophisticated techniques are really but a stage in the evolutionary process that began with ancient man's curiosity about his surroundings. Man has always been interested in the earth and the things on it. Today this interest may be channeled through the discipline of geography, which offers one means of organizing a vast amount of physical and cultural information.

The **Brown Foundations of Geography Series** has been created to facilitate the study of physical, cultural, and methodological geography at the college level. The **Series** is a carefully selected group of titles that covers the wide spectrum of basic geography. While the individual titles are self-contained, collectively they comprise a modern synthesis of major geographical principles. The underlying theme of each book is to foster an awareness of geography as an imaginative, evolving science.

Preface

The present uncertainties characterizing the world of energy supply and demand, have prompted attention to this sphere of activity. Increasing energy consumption, changing preferences for different sources of energy, growing dependence on foreign supplies, and the resulting pressure on national economies and international politics, are causing government and industrial leaders to continually reevaluate their policy decisions. Larger volumes of energy have demanded new types and forms of transportation media. New routes have been established for transporting energy, and old routes expanded. Old and new energy producing regions have expanded while occasionally the exhaustion of the energy resource or a changing market has caused the decline of a once important region.

Around the world, millions of people are striving for an improved level of living, while other millions are tackling the job of cleaning up an environment dirtied in part by energy consumption. Both activities are vital to those involved, but will require an increase in energy consumption.

To provide the energy needed during the rest of this century, new discoveries of the traditional energy resources must be located and developed while, at the same time, research must be undertaken which hopefully will allow for the economic utilization of such energy sources as solar power, geothermal power, and hydrogen fusion.

It is the purpose of this book to provide the reader with a better understanding of the factors involved in the production, transportation, and consumption of energy, and the impact of these activities on the environment. In a book of this length, it is necessary to limit the amount of material, and the author accepts full responsibility for such selection and emphasis.

This book has resulted from the stimulus provided by former professors, colleagues, students, editors, and relatives. The author's interest in the field of energy dates back to the early sixties with the writing of a master's thesis on energy consumption in Europe. This was followed by a doctoral dissertation on the geography of the electric power industry in Michigan. Continued reading and research in the field of energy, and the offering of a geography of energy course for several years, has served to increase the author's interest in this field.

The author wishes to thank Professors George Kish and John D. Nystuen for their guidance in the study of the geography of energy, Richard P. Thiel who early encouraged the writing of this book, and to his wife, Susi, who has sustained this encouragement.

H. Reid Wagstaff

Contents

Tables

Chapter 1

Nonmodern Energy Resources

In varying degrees, man affects the environment when he produces those resources needed to provide energy. The extent and kinds of change imposed on the landscape vary with the scale of the assault for the particular resource. Although significant changes have taken place in past history, man is most concerned today with the impending onslaught demanded by tomorrow's energy needs. This applies not only to those energy resources expected to be consumed by the advanced nations, but also to those by which the underdeveloped nations will seek to improve their level of economic development.

Classification of Energy

Energy may be classified as being animate or inanimate. Human porters, draft animals, and wood are examples of animate energy while coal, natural gas, and waterpower are considered to be inanimate. Energy may also be classified as being renewable or nonrenewable. Renewable forms include wood, the wind, the sun, the tides, geothermal steam, waterpower, animals, and human beings. Coal, petroleum, and natural gas are examples of nonrenewable energy sources.

renewable energy

The major advantage of renewable energy is that the sources of energy are renewed by natural means. Waterpower is renewed via the evaporation-precipitation cycle, although the distribution of precipitation varies extremely unevenly throughout the world, and in any given area may vary greatly from year to year. Groundwater, seeping beneath the earth's surface, may come in contact with very hot rocks, thus releasing a flow of natural steam or hot water for possible use by mankind. The changes in the amount of gravitation force exerted on the ocean waters by the daily and monthly changes in the relative positions of the sun, the moon, and the earth, produce tides which occur unequally along the shores of the continents. This tremendous source of energy is used so little today. Atmospheric changes, resulting from the absorption or loss of radiant energy from the sun, produce winds which continually vary in intensity, direction, duration, and location. Even though extreme amounts of energy are concentrated in the winds of tornadoes and hurricanes, man has not yet been able to harness this power source. Limited success has been achieved by using windmills, but in the Netherlands (long identified with wind power) small, efficient, dependable, and more economical diesel engines have all but displaced the Dutch windmill.

The potential energy that might be obtained from the sun is limited by the interruptions brought on by nighttime and cloud cover. Also, there is the expense of conversion equipment. The world's forests reproduce themselves naturally, although in some select areas man speeds up the process by transplanting seedlings.

And, of course, there is man himself. Manpower is used in many nations because it is so cheap and abundant when compared to the alternatives. Of the above-mentioned renewable energy resources, manpower is the most widely utilized form, but waterpower is most significant in providing concentrated energy in the form of electricity for large-scale use.

nonrenewable energy

Coal, petroleum, natural gas, and uranium are the major forms of nonrenewable energy (table 1.1). Once the energy has been produced by burning or fission processes, new sources of energy must be located and used. Since these sources have been produced over geologic time, it is impossible to wait for the creation of new supplies. Thus the search to find new, economically feasible deposits to develop, will continue and research will hopefully improve our utilization of these energy sources in order to extend the supply. In the United States, the search is complicated by the fact that we are no longer just looking for new sources of energy so that we can choose between them; rather, we are striving to locate sufficient energy sources simply to meet the expanding demand in all areas.

Human Beings and Animals

In some nations, the cheapest form of energy available is human labor. The lack of low-cost local or imported energy is one factor which encourages the use of human labor. Others include the desire of the government

TABLE 1.1

World Energy Production, 1969

(quantities in million metric tons of coal equivalent)

	Total Energy	Coal & Lignite	Crude Petroleum	Natural Gas	Hydro & Nuclear Electricity
World	6,526	2,330	2,741	1,307	148
North America	2,141	523	701	864	52
Latin America	413	10	345	50	9
Western Europe	542	394	26	76	46
Western Asia	824	6	803	14	1
Far East	236	137	68	16	15
Oceania	57	52	3	0	3
Africa	383	57	319	4	3
USSR & Satellites	1,929	1,150	476	282	21

SOURCE: *Statistical Papers, Series J, World Energy Supplies,* United Nations 1959-1969.

to save exchange capital by not purchasing all the foreign energy which could be used, while at the same time providing as many jobs as possible for the people. And then, some societies have not yet advanced to the stage of using modern applications of energy in its various forms.

In a primitive society, human energy is always available to do the work involved in hunting and fishing, farming, building shelters for man and animal, and collecting wood for burning. Man's source of energy is thus always with him, ready to perform the desired task. Yet he is sometimes faced with the need to use his energy to gather wood and transport it to the area of use, usually the village of the tribe.

Man early learned the advantage of having animals do some of the work and thus he began to domesticate the animals native to the area, or introduced those which could adapt to physical characteristics of the region. The long list of tasks to which animals have been applied includes examples from around the world, in a variety of environments. In many societies today there still exists situations in which an animal—because of its cost, capacity, and utility—is the best form of energy for the people. But the continued use of animals is under constant threat from new types of machinery, such as the snowmobile in place of the Eskimo dog, or the two-wheeled garden tractor in place of the Asian water buffalo. The following passage illustrates the problems of relying on animal power when the real need is mechanization.

> Horse and mule numbers at that time (1918) were the highest in our history —more than 25 million——but the rate of technological progress had slowed

> down. The availability of good new land had dwindled to insignificance. One-fourth of the harvested crop acreage was being used to produce feed for power animals.
>
> If methods had not been changed, many more horses, more men to work them, and much more land to grow feed for them would be required for today's net agricultural output. The American economy of the 1960's could not be supported by an animal-powered agriculture on our essentially fixed—in fact, slowly shrinking—land base. National progress on all fronts would have been retarded seriously had not agriculture received new forms of power and sources of energy not restricted by biological limitations.
>
> With the adoption of mechanical forms of power in engines, tractors, and electric motors and development of more and more types of adapted equipment to use that power, American agriculture entered a new era of sharply rising productivity.[1]

Animals must be considered in another way as a source of energy for man. In different parts of the world, but especially in those nations in the Middle East and Asia where firewood is scarce and thus expensive, dried animal manure is commonly used for fuel. The manure is usually made into cakes while moist, and after drying, they are ready to be used for cooking, and heating homes. In the early days of exploration and settlement of the Great Plains in the United States, many a campfire and fireplace burned the manure of the bison, usually referred to as buffalo chips. Although this provides a form of energy needed by the people, in many such areas the manure is badly needed to enrich the soil.

Wood

From the earliest of times man has used wood to fuel fires: fires which cooked food, provided heat, light, and even protection from wild animals. Forests at one time covered more than two-fifths of the total land area, excluding the polar regions. Man has since deforested something less than one-third of this original forest, leaving a forested area of 10,127 million acres, or about 30 percent of the land surface area. Generally speaking, the forests of the high latitudes tend to be coniferous trees; those of the low latitudes, tropical hardwoods; and those of the middle latitudes, deciduous hardwoods. Coniferous wood is the favorite of the industry, while the hardwoods are given preference for firewood. Of the total cuttings in the mid-sixties, only 17 percent of the softwood, mainly conifers, was used as firewood, whereas 72 percent of the hardwood was burned as fuel. On a world basis, 45 percent of the felled timber was used as firewood.

1. *Yearbook of Agriculture, 1960,* (Washington, D.C.: U.S. Department of Agriculture), pp. 4-5.

Wood is still an important source of energy for people living in a variety of environments. Since it is a renewable resource, a continual harvest is possible so long as overcutting does not persist. In the higher latitudes, cutting is often done during the winter season by fishermen or farmers who need a supplemental income. The speed of felling trees and cutting the logs to proper lengths has been increased by the addition of power saws to the basic equipment of axes and handsaws. Easy transportation to market, especially the floating of logs downstream, and the lack of a need for expensive storage facilities, gave wood certain advantages in many underdeveloped nations. Indeed, various railroads and river steamers around the world still rely on wood, purchased along their routes, to fire their boilers.

In his theory, first published in 1826, Von Thünen hypothesized that the market price less transport costs determined land rent, and thus the spatial pattern of land use. In his explanatory diagram, the different land uses were shown by concentric circles centered on the urban area. The second closest circle to the city was forested land, with crop and pastureland further away. The reason for such a location was the reliance of the city dwellers on wood for heating and cooking. Since the transportation of wood was expensive due to its bulk, which also limited the amount that could be stored at each house, the woodlots became concentrated relatively close to the market. Even today, although wood has long since lost its place as the energy source for heating and cooking, many cities and towns in western Europe are characterized by the remnants of these woodlots which today serve as parks and national forests.

From the first experience of the new settlers in the United States, wood was the major source of energy. It was available in abundant quantities due to the dense forests which covered so much of the eastern half of the nation. In order to prepare farmland, the trees would be cut down, thereby providing the farmer with a huge supply of wood for cooking and heating. In the trans-Appalachian settlement, where the Township and Range Survey System gave land division its checkerboard pattern, farmers often preserved part of the forest on the poorest or most distant piece of their land as a source of both firewood and lumber. Even as industrialization began to take place, wood retained its importance as a fuel, although slowly it was replaced by other fuels in the city because its cost was comparatively greater and partly because it required greater space for storage.

Energy in wood is released through the combustion of two main elements contained in it—carbon and hydrogen. There is a greater proportion of carbon to hydrogen in all woods but this ratio varies from one wood to another. The greater the percentage of carbon, the more heat that will be released, although hydrogen flames temporarily give off considerable heat. However, for most heating purposes, it is the bed of glowing carbon that is most desirable.

There are certain impediments to the burning of wood. One is the presence of the natural sap. Wood left to stand for a period of time (thus allowing the wood to dry out) is called "seasoned" wood. Another problem is that wood easily absorbs moisture from the air and other sources. To be completely free of moisture, wood must be kiln dried. When wood is either "green"—not seasoned—or wet, its ability to give off heat is reduced. This is because the water passes off as steam, along with the other products of combustion, and carries with it, in a latent form, much of the heat which was created by the burning of carbon and hydrogen. Thus the importance of seasoning fuel wood and keeping it under cover is obvious if one wishes to obtain the greatest benefit.

As a general rule of thumb, the more dense the type of wood, the greater is its value as a fuel. In the United States, the most valuable fuel wood is obtained from the hickory tree family. Even with its nine species, however, its supply is very small. The second best trees are the oaks, but among the fifty or so species, while some are almost as good as the hickories, many others are poor. The white oak gives off the most heat. Ash, maple (especially the sugar maple and black maple), and the black or sweet birch complete the list of the most desirable woods. Trees of fairly high quality include the beech, crab apple, dogwood, and honey locust; while the elm, black walnut, and black locust are of lower value. Chestnut wood, while it produces a fair amount of heat is often objectionable because of the sparks it throws off. Near the bottom of the list can be found the cottonwoods, poplars, willows, and linden trees.

The conifers provide large quantities of firewood, especially in the western and southeastern United States. These woods ignite easily because of the large amount of resin present. However, the proportion of hydrogen to carbon is about the same, and thus the fire soon burns out after a period of hot flames, leaving behind little ash. If converted to charcoal, its quality is poor due to softness and low heat content.

Charcoal

Charcoal is wood which has been freed from its hydrogen and other volatile contents by heat. This heating process takes place in a kiln, which may or may not capture its by-products, the most important of which is wood alcohol. Charcoal is a desirable product because it is almost pure carbon, burning freely and steadily. This conversion of wood to charcoal also reduces the bulk of firewood, concentrating more BTU's in smaller, lighter chunks.

Charcoal can be made in rather crude kilns fashioned in or near the forests. The wood is first cut into pieces, perhaps one-to-two-feet in height. These pieces are then stacked on end in a dome-shaped pile on the ground, after which sod and dirt are used to cover the wood, except for openings left at the top and around the base. The wood is then burned, beginning at the top. Vents, dug close to the top of the kiln, allow air to circulate around the top layers of wood. The first smoke which escapes is black, but soon it becomes transparent. When this changes to a bright blue, it indicates that carbonization is close to the vent. That vent is sealed and a new one dug into the sod and earth at a lower level, thus allowing a new layer of wood to be burned. This process is repeated until all the wood has been burned. The result of such crude practices is the complete waste of all by-products, and charcoal which contains pieces of charred wood and dirt. When the burning is completed, the mound is covered with moist earth and allowed to cool for several days.

This process will reduce the weight of some of the best charcoal woods to about 20 percent. Such weight reductions make it possible for charcoal to be shipped long distances in the United States to feed the backyard barbecues. Its light weight makes it easy to carry from the supermarket to the car, house, apartment, or picnic grounds. Its small bulk, high-heat output, and relative cleanness have encouraged its use in this country.

The modern beehive kiln has a dome or hemispherical shape, and is made mainly of bricks. The opening at the apex of the kiln is used as a vent as well as the opening through which fresh wood is introduced. At the base of the kiln, openings provide air to kindle the fire, with one large opening existing for the purpose of allowing the removal of charcoal. Charcoal made from softwood tends to be dirty to handle and burns more quickly. The best charcoal, made from hardwood, when shaken together will produce an almost metallic tinkle.

In the United States, it is estimated that wood reached its peak in 1850 when it supplied 90 percent of the energy needs of the nation. But with the growing popularity of coal, its position quickly slipped to only 21 percent in 1900, and to less than 1 percent today. Statistics on wood consumption are difficult to obtain and usually are derived from educated guesses for large areas of the world. Nonetheless, on a world basis, it is estimated that wood produced about 10 percent of the world's energy in the mid-sixties, a percentage which will continue to decline.

In the early development of the iron and steel industry in the United States, Pittsburgh became important because of the initial supply of iron ore in the area, and the abundant forests, which served as the source for making charcoal, the major fuel used to refine the iron ore. With the discovery of the superiority of coking coal in the iron and steel industry, the use of charcoal was abandoned. It is interesting to note that, in recent years, the Chinese have resorted to charcoal for refining iron ore in their backyard ovens.

Charcoal has historically been an important fuel in many countries. As stated earlier with regard to the United States, charcoal's cleanness, light weight, little ash residue, and high-energy output make it a natural fuel for long-distance movement, especially in underdeveloped nations. The charcoal is easy to ignite, provides a steady heat, and is as effective in small fires for cooking as in large-scale applications, such as iron ore smelting. China's increase in iron and steel production has been in part due to the widespread use of small, backyard iron ore smelters, fueled by charcoal. In many Asian nations where the population has settled and farmed the flat lowlands, trees cut in the highlands and mountains are converted into charcoal which can then be brought down to the cities and villages where part is swapped for the agricultural products needed by the forest peoples.

Today, in the United States, the demand for charcoal has increased its cost, in part due to the limited production. Because of the price increase, many firms have found it profitable to manufacture charcoal briquettes. These are sometimes pure charcoal but some are made by compacting a mixture of wood wastes, and even coal particles. Furthermore, good hardwood suitable for charcoal production is becoming more scarce.

The whole topic of harvesting our forest products is coming under attack. Many groups oppose the method by which trees are harvested—the method called *clearcutting,* in particular. By this technique, every tree in an area of many acres to a few square miles is cut down. This economizes the effort of harvesting trees. But the effect of such practices is to expose the soil to erosion. Thus, *selective cutting,* the practice wherein only the mature trees are cut down, is to be preferred because with its use the beauty of the forest is retained while the ecology of the area is relatively slightly disturbed. Some areas, lying relatively flat and so being in minimal danger from running water erosion, can be clear-cut without major problems.

Beyond the boundaries of the United States, in regions such as tropical America, charcoal remains the most common household fuel. Although more modern fuels will continue to make inroads into such regions, the availability and thus cheapness of wood for making into charcoal, will enable it to retain an important share of the market.

Peat

Peat is the earliest stage in the development of coal. It is composed of partially carbonated, decayed vegetation which was originally deposited in water, and sometimes covered by silt. Peat can be mined or "cut" from a peat bog or "peat bank," which is usually found near the surface of the ground.

The basic task in producing hand-won peat or turf, is to cut it into chunks or sods with an ax or somewhat similar tool. It is then stacked nearby for the initial drying process during which the water content is reduced to about 35 percent by evaporation. Since the relatively dry air is the mechanism for drying the peat, it is harvested only in the spring and summer months. When either season is wetter than usual, the result is a sharp decline in peat production. After transporting the sods to the home, store, or power plant by wheelbarrow, horse and wagon, or truck, it is restacked for final drying before use. Most of this hand-won turf is used by the farmers who dig it, and turbary[2] rights are an important aspect of land sales and tenancy agreements.

Since the 1930s, a state-directed agency in Ireland has encouraged the development of machine turf production and the bulk of this output goes to the several electric power plants built after World War II and designed for this fuel. Giant bucket dredgers which work sod faces up to twelve feet in depth and rotating, spiked drums which cut about three-eighths of an inch into large flat bogs are producing huge quantities of sods and briquettes. In some cases, light railroads transport the peat to the power plants. The life expectancy of these commercially worked bogs is about thirty years, whereas the shallower, noncommercial bogs may last another century.

On a worldwide basis, peat production represents less than 1 percent of total energy production, with little prospect for expansion. Even in Ireland, if other jobs were to become available and imported fuels were to remain somewhat cheaper than peat, peat production would decline rapidly.

The production of peat represents an interesting phenomena in an age of ecological concern. Peat bogs are a definite characteristic of the cultural landscape and, as such, their relatively small size and apparent inoffensiveness to the people living and traveling in Ireland, has caused little concern over their existence. One can logically expect their existence to continue far into the future.

2. "A turbary right is an easement under English law to dig turf or peat on a common or on another's land." (By permission. From Webster's Third New International Dictionary © 1971 by G. & C. Merriam Co., Publishers of the Merriam-Webster Dictionaries.)

Chapter 2

Transporting Firewood

Man has always had energy, in one form or another, close at hand. The caveman relied on his muscles, and the ability to transport firewood from place to place, and put it to service for his benefit. But the capacity of such human power was limited, and soon the importance of domesticated animals became evident. This gave man an alternative, and he found new applications for such use, freeing him from some physical exertion. He could also move this energy source with him, wherever the task might be. However, the continued development of man, especially as he proceeded into the industrial and technological revolutions, has affected the lives of people living in underdeveloped countries or rural areas. Modern forms of energy are replacing the unique and usually backbreaking ways of doing work. For example, in Hong Kong, the government no longer issues licenses for men who wish to pull pedicabs and each year witnesses the loss of some of the older men. Therefore, pedicabs, long a way of life and an important form of transportation, will disappear from the Hong Kong scene. In many Asian rice paddies, modified two-wheeled garden tractors have proven to be faster and more versatile than the traditional draft animals. In Venice, the speed and efficiency of outboard motors makes it appear likely that they will displace the romantic gondolier and his gondola. Each year, then, witnesses a wider use of modern energy resources because man continually seeks to find easier, more efficient and economic ways to accomplish his tasks.

Wood is a high-bulk, low-value product. The transportation of wood for long distances forces its cost up, often to the point of encouraging and allowing the substitution of other energy sources. The availability of low-cost transportation also determines whether or not certain stands of timber will be cut for fuel. In the exploitation of the taiga forests of Canada and Scandinavia, much wood remains uncut due simply to a lack of low-cost transportation. However, large acreages are harvested where the logs can be floated cheaply and easily down a river, or carried by rail, to market.

In the United States, few homes still rely on wood as the energy source. Those that do consume firewood, reflect the cultural preference for it as a primary or secondary heat source, or the economic advantage due to its relative abundance and low cost.

Timber can be moved by wagon, train, ship, or raft. In the northern Great Lakes states, countless wagons were constructed to move timber south to the urban markets. Rafts were also used to move logs across the Great Lakes, and from the Pacific Northwest to markets as far south as San Diego. Trains and ships still move large amounts of timber, but much of this is cut into lumber, with only the slabwood available for firewood. In other parts of the world, however, wood is still used to fuel old riverboats and steam engines, and is carried by watercraft, railroad, truck, pack animal, and even by human porter. The type of carrier in large part depends upon the level of economic development of the nation or region.

Since the end of the Second World War, the United States has experienced great expansion into the rural land surrounding major cities and towns. This included the construction of many single-family houses, often with a wood-burning fireplace in the living or family room. The result has been a continued increasing demand for firewood, firewood that could no longer be locally cut because of the density of houses in the area. Nonetheless, in most suburban areas, it was feasible to take a drive in the family car to pick up or purchase firewood. In the last decade or so, this situation has changed. The demand for wood has so increased, and the population in general so willing to pay higher prices, that now lumber mills that once gave away slabwood to anyone willing to take it are now selling this scrap wood at the mills. It is also tied in bundles, to be sold miles away through supermarkets, convenience markets, and even gas stations. With such a market, some people find it profitable to cut timber for firewood and drive into the urban areas, selling it door-to-door. In grass or desert lands, such distances can be very great, causing the price of hardwood to exceed fifty dollars per cord. The continued increase in the cost of firewood, coupled with a decreasing supply, suggests that fewer fireplaces will be built, per household, in the coming decades. And the tantalizing smell of wood smoke on a cold, winter's night, will become increasingly rare.

Chapter **3**

The Mining of Coal

Merriam-Webster defines coal as being "a black or brownish black solid combustible mineral substance formed by the partial decomposition of vegetable matter without free access of air and under the influence of moisture and in many cases increased pressure and temperature, the substance being widely used as a natural fuel and containing carbon, hydrogen, oxygen, nitrogen, and sulfur as well as inorganic constituents that are left behind as ash after burning."[1] Many others define coal as that resource whose presence in a significant deposit within the country's boundaries, represents one of the traditional needs for any nation hoping to build a strong industrial base. Historically, coal is acknowledged as the fuel which made possible the industrial revolution.

Formation

Coal is formed from the remains of woody and herbaceous swamp plants which were covered first by water and eventually by one or more layers of sediment. With the passage of time, the thick vegetation of these swamps becomes peat, protected from complete decay by its submergence in water. As heat and pressure are applied to the partially decayed vegetation by the addition of more sediment and rock above it, coal can be formed.

The lowest form of coal is lignite or brown coal, which is very friable and has a low heat rate per unit of weight. Additional heat and pressure can produce bituminous or soft coal, which is the type of coal used in the largest quantities and is also the source of coke. The highest form of coal,

1. (By permission. From Webster's Third New International Dictionary © 1971 by G. & C. Merriam Co., Publishers of the Merriam-Webster Dictionaries.)

produced in areas which have experienced very high temperatures and pressure due to the folding and faulting of the rock, is anthracite or hard coal. This coal is usually the most expensive to mine but historically met an important need—the demand for a slow-burning but hot coal which had a minimum of obnoxious odors, coal dust, and residue ash. Almost all of this high-grade coal mined in the United States has come from the anthracite district of northeast Pennsylvania. However, when the American homeowner began to choose oil or natural gas as his space-heating fuel, this region suffered an economic depression since no other market existed for this product on such a large scale.

coal production in the united states

The Appalachian Field has long been the most important coal field in the United States, accounting for the high ranking of Pennsylvania, West Virginia, Ohio, and in part Kentucky, over the years (table 3.1). The Eastern Interior Field is next in importance as evidenced by the coal mined in Illinois, Indiana, and in part Kentucky. The Rocky Mountain Field is just beginning to develop its extensive reserves and thus, in 1970, only Wyoming produced enough coal to rank in the top ten states.

world coal production

In 1970, the United States regained its ranking as the world's leading coal producer, narrowly surpassing the U.S.S.R. (table 3.2). The increased energy demand, especially for electricity produced in part by coal-burning power plants, pushed the American production up, while production

TABLE 3.1

The Ten Leading Coal Producing States 1960-70

(thousands of tons)

	1960	1970
1. West Virginia	118,944	144,072
2. Kentucky	66,847	125,305
3. Pennsylvania	65,425	80,491
4. Illinois	45,977	65,119
5. Ohio	33,957	55,351
6. Virginia	27,838	35,016
7. Indiana	15,538	22,263
8. Alabama	13,011	20,560
9. Tennessee	5,931	8,237
10. Wyoming	2,024	7,222
Total	434,329	612,661

SOURCE: *Statistical Abstract of the United States*, (Washington, D.C.: U.S. Department of Commerce, 1972).

TABLE 3.2

The Ten Leading Nations in the Production of Bituminous Coals, Anthracite, and Lignite, 1970

(tons)

Rank	Country	Production
1.	United States	596,969,000
2.	U.S.S.R.	595,553,000
3.	China	400,000,000
4.	East Germany	288,672,000
5.	West Germany	230,186,000
6.	Poland	190,553,000
7.	United Kingdom	148,318,000
8.	Czechoslovakia	121,073,000
9.	India	85,141,000
10.	South Africa	58,350,000
	World Total	3,096,807,000

SOURCE: *Minerals Yearbook, Vol. I, Metals, Minerals, and Fuels,* (Washington, D.C.: U.S. Department of the Interior, 1971).

slumped somewhat in the U.S.S.R. This is even more significant when it is understood that most of America's production is bituminous coal with a small amount of anthracite, whereas the Russian production includes a large percentage of lignite. The figure for China is approximate, although it is well known that China does possess some outstanding fields and very large reserves. It is also interesting to note that practically all of East Germany's production is lignite, while that of West Germany is mostly bituminous.

world coal reserves

A glance at the statistics (table 3.3) suggest that there will be little change in the world rankings of coal producing countries. The United States contains almost 50 percent of the known coal reserves of the world, and when this amount is added to the reserves of Asia, including the U.S.S.R., 85 percent of the world's coal reserves are included. Not only does this indicate the coming depletion of the famous coal fields of Europe which began the industrial revolution, but it also indicates the concentration of coal deposits in the Northern Hemisphere.

Factors of Production

Throughout the world, many coal deposits lie undisturbed for a variety of reasons, most of which relate to the cost of producing and delivering the coal to market. One factor is the depth of the deposit beneath the surface of the ground. The deeper the coal, the greater the cost for lifting it to the surface. Thus in the United States, deep coal seams exist in the Appalachian Field, for example, which remain unmined simply because other supplies of coal are cheaper to mine right now. After the shallower seams are mined, the deeper seams will be analyzed to see if they can be removed economically.

In the United States, the majority of the underground mines are less than 1500 feet beneath the earth's surface, while some foreign countries, whose supply of coal is much more limited, have mines more than twice as deep. In strip or surface mines, the thickness of the overburden that can be removed is directly related to the thickness of the seam. The ratio is roughly thirty to one. That is, for every foot of thickness of the coal seam, thirty feet of overburden can be removed, although at the present about 150 feet appears to be the temporary economic limit.

Another production factor is the interplay between distance to market and intervening opportunity. If production costs are about the same at two different mines producing the same grade of coal for the same market, but one mine is located hundreds of miles closer to the market, the closer mine should be able to dominate that market. Intervening opportunity can be explained by using the previous example and stating that the more distant mine was once the sole supplier of the market which paid the price asked by the mine's management. Then the second and closer mine opened its coal operation. Because it could produce the same grade of coal as the more distant mine and sell it at a lower price (due to the shorter haul required for moving the coal to the market), it soon dominated the coal market in that city. Thus the management of this mine was successful in part because its location represented an intervening opportunity in the sale of coal to this city. The more distant mine's management must now decide whether or not to shut down its operations, seek another market, or determine if there are ways and means of reducing its mining and transportation costs so that it can compete with the new mine.

Another variation is that in which the closer mine exists and is in full operation supplying the city with the coal. The coal is known to exist at the site of the more distant mine in the example above. But in this case, the market is limited and therefore the more distant mine does not warrant development.

Examples such as the above are representative of the types of competition and changing market conditions which constantly face the coal mining industry, as well as the petroleum and natural gas industries.

Still another production factor is the degree of mechanization used in mining the coal. Strip or surface mining usually has the highest production per worker due to the use of huge machines to clear away the over-

TABLE 3.3

Estimated Remaining Coal Reserves of the World

(millions of tons)

Region: Nation	Anthracite, Bituminous, and Sub-bituminous Coal	Lignite and Brown Coal	Total All Coal	Percentage of World Total
Asia				
U.S.S.R.	1,099,796	222,604	1,322,400	19.9
China	1,114,122	661	1,114,783	16.7
India	68,795	560	69,354	1.0
Japan	10,906	284	11,191	0.2
Others	5,192	4,196	9,388	0.1
Total	2,298,811	228,306	2,527,116	37.9
North America				
United States	2,288,810	921,250	3,210,060	48.2
Canada	68,844	26,944	95,788	1.4
Mexico	4,745	–	4,745	0.1
Total	2,362,399	948,194	3,310,593	49.7
Europe				
Germany	247,179	68,324	315,503	4.8
United Kingdom	188,096	–	188,096	2.8
Poland	88,160	20	88,180	1.3
Czechoslovakia	7,108	13,775	20,883	0.3
France	13,541	474	14,015	0.2
Belgium	6,599	–	6,599	0.1
Netherlands	3,747	–	3,747	0.1
Others	2,889	24,414	27,414	0.4
Total	557,319	107,006	664,325	10.0

Africa				
Republic of So. Africa	74,936	—	74,936	1.1
Others	1,818	220	2,039	—
Total	76,754	220	76,975	1.1
Australasia				
Australia	18,514	45,182	64,670	1.0
Others	109	865	974	—
Total	18,623	46,047	64,670	1.0
South and Central America				
Columbia	13,224	—	13,224	0.2
Venezuela	3,381	0	3,381	0.1
Others	3,919	280	4,199	0.1
Total	20,524	280	20,804	0.3
World	5,334,430	1,330,053	6,664,483	100.0

SOURCE: Compiled from Geological Survey Bulletin 1136, January 1, 1960. U.S. data from Geological Survey 1275, January 1, 1967.

burden and dig out the coal. Although the cost of such machinery is very great, the small labor requirement with a corresponding high productivity per worker combine to produce coal at the lowest cost. Underground mines have a greater investment in the shafts and tunnels and incur greater costs in lifting the coal to the surface and operating the lighting, ventilation, and transportation systems, but mechanization can greatly increase productivity per worker. The accompanying statistics point out the advantages of surface mining and more modern underground mines in the United States when compared to European coal production (table 3.4).

TABLE 3.4

Output Per Man Per Day at United States and European Coal Mines, 1938-67

(net tons)

	1938	1947	1957	1967
United States				
Underground	4.60	5.49	8.91	15.07
Strip	15.00	15.93	21.64	35.17
Average all mines	4.89	6.42	10.59	19.17
United Kingdom	1.67	1.61	1.76	3.15
Belgium	1.20	0.95	1.27	2.46
France	1.35	1.05	1.85	2.32
Netherlands	2.61	1.81	1.65	2.61
West Germany	---	1.32	1.75	3.60
Poland	---	1.86	1.78	2.56

SOURCE: *Bituminous Coal Facts*, (Washington, D.C.: National Coal Association, 1970).

An interesting example of the interplay of different factors on coal production relates to the Rocky Mountain Province. It has long been accepted that its major disadvantage, distance to market plus the effect of intervening opportunity as represented by the Western and Eastern Interior Fields and the Appalachian Field, made Rocky Mountain coal too expensive when delivered to the Manufacturing Belt.[2] Now, because of the concern over air pollution in the major metropolitan areas, solutions are being sought which will reduce the amount of pollutants released into the air from power plant stacks. One partial solution is to switch to the low-sulfur coals that exist in huge quantities in the Rocky Mountain Field. The coal may cost more than that obtained from the former source due to the long transportation haul, but the low sulfur content in many cases

2. The Manufacturing Belt is approximated by the box formed when a line is drawn connecting the cities of Milwaukee, St. Louis, Baltimore, and Boston (then back to Milwaukee). This region contains the greatest concentration of manufacturing activities in the United States.

meets the limits set by law and is cheaper than alternative fuels in some cases. Thus stiffer air pollution laws, the expense of pollution control equipment, and the scarcity and expense of alternative fuels may well encourage the development of coal deposits long deemed uneconomic to mine.

Mining Techniques

The two dominant coal mining methods are underground mining (shaft and tunnel for the most part) and strip or surface mining. The two variations of underground mining are known as the room-and-pillar, and longwall mining. The room-and-pillar method is used extensively in the United States because of the relative abundance of coal. Large blocks or pillars are left to support the rock ceiling. This reduces the cost of mining but is very wasteful, leaving behind perhaps as much as 50 percent of the coal. The other method, known as longwall mining, is used extensively in Europe, where it is necessary to remove a greater percentage of the coal, due to that continent's smaller total reserves, and the generally greater depths at which the coal seams lie. In this method, coal is removed along a continuous face, after which many supports of heavy timber are erected, or roof bolts inserted, to support the ceiling of the mine.

Strip mining is dependent upon the use of huge power shovels or draglines. Although these machines are very costly, such mechanization has resulted in the ability to produce coal at a very low cost. The evolution of ever larger shovels and draglines has allowed strip mining to go deeper, thus making it possible to dig coal which previously was too expensive for strip mining and physically and/or economically impossible for underground mining. In the United States about one-third of the coal produced is mined from strip mines as compared with the 63 percent produced from underground mines. On a world basis, production from underground methods is even more dominating.

In the statistics cited earlier, it was shown how strip mining (or surface mining, as the industry prefers it today) has allowed the American industry to produce large amounts of coal at a very low cost via intensively utilized, large-scale machinery and equipment. Surface mining has certainly left a bad impression in Kentucky where many square miles of land have essentially been destroyed. To better understand the situation, it should be explained that there are two main types of surface mining: (1) strip mining as it is carried out in a large area on relatively flat land, and (2) contour strip mining. Of the two, contour strip mining, partially because it is usually carried out on a steeply sloped land, is most devastating to the landscape.

In conventional strip mining, the land is relatively flat and the disturbance that does take place exposes deeper layers of the earth's surface. This often includes angular rock pieces created by blasting and by the power shovels. Such overturning of rock material often exposes material that is devoid of much fertility to encourage and support revegetation. But experience thus far has shown that in humid areas, with some grading to achieve a kind of order out of the otherwise sawtooth profile, this overturning of the material can create a soil material that is more productive for some vegetation types, particularly in trees, than the non-strip-mined land. In arid climates, the problem is one of trying to reseed the land with native or introduced vegetation, and hoping that the warm season will not be unseasonably dry. In one such area, located on an American Indian reservation, the growth that took place on the test plot was quickly consumed by grazing sheep, which had been let in through holes cut in the fence. Thus accurate measurements and comparisons of growth on strip-mined and nonmined land in the area was impossible. Illinois exemplifies the good and bad effects of strip mining in a humid region of relatively flat land. It is possible to drive about that state and see huge angular mounds of overburden, rather devoid of vegetation, containing pools of stagnant water, in part caused by the interruption of the natural surface drainage system. It is indeed a spoiled land.

Using examples such as these, it is easy to understand why some groups and individuals have raised the cry to outlaw strip mining in the United States. On the other hand, one can also find examples of land in Illinois which has been reclaimed after the coal was mined so as to benefit the people living in the area, and those at some distance from the area. In these examples, a conscious effort was made by the coal companies to turn the land into productive use. Here are some of the results. The reforested and regrassed ragged topography of the strip-mined landscape became a good habitat for deer, game birds, and other small game such as rabbits, much to the joy of hunters. The last ditch or trench that is cut is left open because of the lack of overburden to fill it in. The result can be the creation of long but narrow lakes suitable for providing recreation foci for the surrounding area. Swimming, boating, and even water skiing can then be available to people who live in areas distant from lakes. Once stocked with game fish, such lakes become a fisherman's delight. In one instance, the creation of such a body of water resulted in a reservoir, which provided the people of a nearby small town with a sure supply of water for municipal use—and also enabled the residents to receive lower fire insurance rates because of the improved ability of the fire department to combat fires with the new, nearby water source.

In a more conventional sense, this mined land can be planted to grasses and trees, creating an environment suitable for the grazing of animals. Although one could argue that it is better to save all land in its original state, the need for energy resources, coal in particular, must overrule this utopian approach. Instead, choices should be made to allow for the strip mining of those areas deemed to present the least value to the public today with the understanding that the land would be reclaimed for certain, predetermined uses.

On the dark side of the ledger is the result of strip mining, properly contour strip mining, which has so desecrated huge areas in Appalachia. This form of mining can be applied when the coal lies in a seam which has outcropped on the hillsides, due to the erosion of the land by running water. To remove the coal, large shovels follow along the side of the hill, removing the overburden and dropping that on the lower slope. The coal is then removed from this "terrace" by shovel and truck, and auger equipment is used to remove even more coal from beneath the high wall. The result is the creation of a landscape dominated by much rock and little soil-building material, and having a steep slope. Here, whether in a humid climate or not, the challenge of reclamation is far greater. The heavy rainfall of the Appalachian region tends to wash the finer particles downslope very quickly, continually baring more large infertile rock material. Revegetation is extremely difficult under these conditions. Furthermore, contour strip mining tends to isolate ecosystems. Consider for the moment a situation where the strip mining may follow a coal seam all around a large hill. The result is the creation of a "no-man's-land" which not only may remain void of vegetation, but will create a physical barrier to the movement of wildlife, since the high wall may be fifty or more feet in height. And too, the heavy rainfall, coupled with the steep slope, increases the probability that mine acid will pollute the streams and rivers, to the detriment of those using these waterways. If strip mining is to be limited, it would seem that the practice of contour strip mining should first be outlawed.

Environmental Impact of Underground Coal Mining

Although there are many advantages or benefits that can be claimed for strip mining on level land, it is more difficult to cite any advantages to underground mining other than the fact that it may be more ecologically desirable than the surface disturbance caused by strip mining. This discussion assumes that we are ignoring the economic benefits of coal mining.

Mine Acid

One of the unfortunate by-products of coal mining is the production of mine acid. Although this is in part accomplished by nature without the help of man, simply by groundwater coming in contact with the coal deposits, coal mines have a way of intensifying the production and run-off of such acid. Tunnels often intercept groundwater, causing an accumulation of water within the mine. If this does not drain out naturally, it has to be pumped from the mine, where it is usually released into a nearby watershed. The acid is formed when the groundwater comes in contact with iron pyrites. This contact results in the partial deposition of iron particles into the water, and the creation of sulfuric acid by the solution of the sulfates in the iron pyrites. When this polluted water joins that running in the stream or river, it upsets the ecosystem. Although the sulfuric acid is highly diluted, it still can kill some of the waterlife, often beginning with the best known sport fish. The areas suffering most from this problem are located in the states of Pennsylvania, Ohio, West Virginia, Kentucky, Illinois, Mississippi, and Tennessee. Fortunately, the heavy spring rainfall, coupled with the run-off from the snowmelt, work to effect an annual flushing out of the watersheds each spring. How much worse the problem would be without this assistance from nature! Other areas having mine acid include the coal mining areas of the British Midlands, France, West and East Germany, and Poland.

We can expect this problem to decrease somewhat in the coming years. Most states now have laws which govern the intentional or nonintentional release of mine acid, but unfortunately these laws mainly apply to new mines. With the continued expansion of coal mining, the additional mine acid, which would have been acid to the watersheds, will be largely controlled. But a major source of this pollutant is from the old coal mines, many of which have been closed or abandoned. A major question becomes one of trying to determine who will pay the cost for curing [the problem created by] these old mines.

Subsidence

Frequently, the removal of coal from beneath the ground's surface will lead to a sagging of the rock layers above. In the United States, the use of the room-and-pillar mining method has worked to keep this problem to a minimum. Furthermore, backfilling with waste material from the mining operation—the rock particles which are brought up with the coal and removed during the washing operation—can also keep the subsidence to a minimum. With greater pressure on the coal companies to produce enough coal at a profit, more longwall mining may take place in this country, leading to a greater chance of subsidence. Subsidence can affect drainage systems, transportation networks, and the inhabitability of buildings. In Germany, in an area with subsidence problems, one river bridge was recently constructed with hydraulic jacks at the main support points. Thus, as the land beneath the bridge changes elevations, the bridge can be maintained at level by simply operating the appropriate jacks.

Flooding

In coal mining communities, tailings piles are a visible feature on the horizon. These man-made mountains are composed of the waste material that is separated from the coal at the processing plants. Since it is a waste problem, it is usually stacked in areas of little value to the coal mining operation. Sometimes it is used to fill valleys, and on occasion may create a reservoir during times of unusually heavy run-off of precipitation. The result may be the sudden release of a wall of water dashing downstream to inflict rather severe damage. Or it may be the saturation of the tailings pile itself which, when sufficiently lubricated, flows like mud downslope. This is most serious where such piles have been created close to homes, public buildings, and transportation arteries. If vegetation could be grown on these tailings piles, it would decrease the problems of flooding, erosion, mudflows, dust, and the eyesore that such mounds create. Unfortunately, such material contains little fertility for vegetation growth, and the challenge remains. Perhaps experiments using processed sewage sludge from urban areas to form an initial base for growth will be successful and the future will witness the greening of these waste piles.

Chapter 4

Carrying Coal to Market

Peat

Peat contains a small amount of energy per ton. As a result, peat is moved short distances, often by very small-scale means. In Ireland, peat is carried to the homes in wheelbarrows, dog and pony carts, or horse and wagons. At the largest commercial pits, trucks and light railroads are used to haul the peat to thermal power plants. Because of its very limited use and small yearly tonnage, the transportation of peat produces little environmental concern.

Coal

Of the major transporters of coal—railroad, ship, barge, truck, and pipeline—the railroad is the most significant medium throughout the world. In the United States, more than 75 percent of the coal is moved by trains (table 4.1). East of the Mississippi River, the hauling of coal is the major source of income for the majority of railroads. Thus, railroad management has always tended to keep a sharp eye on this major activity.

Rates charged by the railroads for moving coal and other commodities are controlled by the Interstate Commerce Commission. This was exemplified by established rates per carload, regardless of how many cars or trainloads were involved. Through the Second World War, the railroads were able to retain their competitive advantage with other transportation media. However, after the war, the competition began upgrading their equipment, with the assistance of publicly financed projects such as interstate highways, navigation projects, and port expansion. Increasing costs for labor and material, as well as inflation, continued to handicap the railroads, sometimes to the point of losing old business, or failing to obtain

TABLE 4.1

How Coal Was Transported from United States Coal Mines, 1939-68

(thousands of net tons)

Year	Loaded at Mine for Shipment By Rail	Loaded at Mine for Shipment By Water	Shipped by Motor Vehicle	Used at Mine	Total Production
1939	331,190	22,229	29,534	11,902	394,855
1948	498,194	26,735	58,260	16,329	599,518
1958	305,642	43,899	50,605	10,300	410,446
1968	396,443	66,885	61,753	20,165	545,245

SOURCE: *Bituminous Coal Facts*, (Washington, D.C.: National Coal Association, 1970).

new contracts for moving coal and other commodities. Clearly the railroads needed some form of relief.

the transportation act

In 1958, the Congress of the United States passed the Transportation Act of 1958, otherwise known as the Smathers Bill. Among other things, this bill allowed the railroads to experiment with new pricing procedures under the watchful eye of the Interstate Commerce Commission. The result was the introduction of multi-car rates, trainload rates, annual volume rates, and other such pricing innovations that, for the first time in a hundred years, introduced some cost-price relationships to railroad rate-making.

Unit Trains

The most important development to come from this new legislation, from the standpoint of transporting coal, was the lower rates made possible by unit trains. Unit trains are composed of a string of 85 to 100 cars carrying coal from its source to a destination, such as a power plant. Such trains often have dedicated power, that is, engines which pull the cars for the entire trip, without the expense in time and money of sorting cars in a marshalling yard, or awaiting the availability of engines from the next railroad, to continue the trip.

The unit train approach has not only given a tariff break to the large-scale consumer of coal, but has enabled the railroads to make more efficient use of their rolling stock. For example, coal cars, which served the Trenton Channel and River Rouge Power Plants, located on the Detroit River, just south of Detroit, Michigan, needed an average of fifteen to thirty days to make the round trip to the coal mines in southeast Ohio, and back. The same cars, shuttling back and forth as a unit train, make the round trip in seventy-two hours, producing an improvement in efficiency of at least 500 percent. Such efficiency allows the railroads to reduce their rates and still earn a profit.

Savings

In this particular situation, the published figures gave the rate for coal transported from southeast Ohio coal mines to Trenton Channel or River Rouge as $3.61 per ton. Unit train deliveries reduced this rate to $2.01 per ton. By combining the deliveries of both plants, the rate was further reduced to $1.57 per ton, producing a net savings of more than 50 percent. Because of examples such as this, which took place in the mid-sixties, many other contracts were signed, resulting in unit train deliveries of coal throughout the United States.

Rail transport of coal is almost invariably linked to existing rail lines. Some few examples do exist where new spurs were added, or even a new track laid down with no connection to the nation's existing network. The newest such railroad, which is presently under construction, will carry coal from the Black Mesa mine near Kayenta, Arizona, to the Navaho Power Plant being built on the south side of Lake Powell, on the Colorado River. Its only use will be the moving of unit trainloads of coal.

Lake Captives

Another situation that has been repeated around the world can be exemplified by the St. Clair Power Plant, located near St. Clair, Michigan. Since its completion, it was at first dependent on lake freighter deliveries of coal. Hence it was considered to be a captive of water transportation. After the success of the unit trains at other Michigan power plants, it was decided that a rail spur should be built. Initially this was to insure the plant from being charged higher rates by the freighter lines, who perhaps might be upset over the loss of other business. In reality, it led to the use of unit trains as the least expensive way of supplying the plant with the millions of tons of coal burned each year.

Terminal Costs

Loading of railroad cars, and assembling them into trains, usually involves much time and money. When totalled, they comprise what is known as *terminal costs*. Trains must travel longer distances to offset such costs, which are higher than those for trucks. As a result, short hauls are often dominated by truck or barge traffic, where possible. However, longer coal hauls of one or two hundred miles or more are performed by railroads. To reduce such terminal costs, the coal industry has been developing better ways of loading coal cars. The latest involves pulling the string of cars beneath coal-loading chutes at slow speeds of a couple of miles an hour. High-capacity chutes rapidly fill the cars by this technique which is known as *flood-loading.*

ENVIRONMENTAL IMPACT

It is difficult to separate the factors of rail transportation of coal from all rail transportation, because, as noted earlier, most tracks were built for a variety of uses. The movement of coal by unit and nonunit train certainly produces noise, dust, and traffic congestion at crossings. On the other hand, the transportation of the coal by trucks would be far more offensive, and the feeling today is that a greater movement of freight by railroads—which use less fuel, create less air pollution, and do not need more land for highways—is more desirable.

Trucks

Flexibility and low terminal costs gain the nod in many applications of hauling commodities. In the case of coal, this is especially true where there are several small mines, not close together, that supply coal to a power plant or iron and steel company, for example. The expense of building the railroad, or of operating it, would most likely amount to more than truck transportation. Throughout the Appalachian coal field area, trucks are to be seen carrying loads of coal from the smaller coal mines to large and small customer alike. The newer mines, with their large-scale investment and production, can justify rail transportation. The development of newer, larger trucks, with increased payloads, has given the trucks greater economies of operation. Nevertheless, the truck, if it can compete with the railroads, is usually found serving the smaller producers and carrying coal for distances usually under 100 miles.

ENVIRONMENTAL IMPACT

From an environmental standpoint, truck transportation is not very attractive. Its flexibility allows it to roll on a wide variety of streets, some of which are not capable of supporting either the weight or the intensity of traffic. This leads to the deterioration of roads, including increased dust. The trucks create a real menace in terms of noise, changing relatively quiet neighborhoods, urban or rural, into noise-filled corridors. Of late, greater emphasis is being paid to the air pollution and high energy consumption attributable to such trucks. The major advantages of trucking appear to be in the realm of economic factors, such as the ability to create employment and jobs by working coal deposits too small for rail transport.

Inland Water Barges

Some coal mines are located next to navigable rivers or waterways. In these cases, it costs less to load the coal into river barges and float the coal to its market, if that market is also located along the river. Barges need relatively little power to move huge tonnages because the water is supporting the weight. Therefore, the energy cost is small, and, as long as the waterway is there, the total cost is very low. Part of the reason for the ability of inland barges to offer low-cost transportation is the fact that their waterways, or "right-of-ways," are constructed and/or maintained by the Corps of Engineers at taxpayers' expense. Many port facilities are built and maintained by the barge companies or the coal consumer, but the expense involved is relatively small, due to their simplicity.

In the United States, millions of tons of coal are transported via our inland waterway system. The Mississippi-Missouri Rivers, together with such tributaries as the Illinois, Tennessee, Ohio, Kanawha, and Monongahela, move the bulk of American coal. The low rates of this form of transportation have attracted industries to locate along such waterways, thereby reducing their power costs. On parts of the Ohio and Monongahela, for example, such concentration of industrial users and energy consumption has created rather severe environmental conditions. Expansion of such barge deliveries of coal seem rather slight, assuming the continued existence of other transportation media. Environmental limitations on coal quality burned within metropolitan areas have already affected present and future production of high-sulfur content coal, and this could seriously readjust the flow of coal.

Europe and the U.S.S.R. also rely on river barges to move a considerable portion of their coal. Certainly, the Ruhr Industrial Complex owes a good part of its importance to the navigable waterways which move coal about so easily. In other parts of the world, the lack of navigable waterways accessible to coal mines encourages the use of rail transportation.

Lake Freighters

The demand for coal from the Appalachian Field, by power plants and industries located along the shores of the Great Lakes, at first was met by rail transport. As the level of consumption and the number of consumers increased, it became profitable to transport coal to lake ports, where it was dumped into lake freighters for final delivery by low-cost water transport.

The early vessels used in this trade were simply lake freighters which carried a wide variety of other goods. As the amounts of coal increased, vessels were constructed which are now identified with the special conditions of the Great Lakes. These ships were designed so that they could just squeeze through the locks. Thus, they look somewhat ungainly, rather like barges with small rounded sterns and a shallow, rather broad, bow. But they are economical, and, over the last several decades, the only changes have been in scale, as newer, larger locks have made possible bigger ships. Most of the large lake freighters in use today are about 700 feet long and are designed to navigate the twenty-foot channels. The continued construction of new, larger locks and by-pass canals has opened up the way for the building, or modification of, four outsized U.S. ships, ranging in size from 806 to 1,000 feet in length, and designed to operate in the newer channels which are twenty-seven feet deep. Although these four vessels are being used to haul iron ore pellets, they certainly will open the way for larger coal-carrying vessels.

Types of Freighters

The coal-carrying Great Lakes freighters can be divided into those that are self-unloaders and those which require shore-mounted equipment to unload their cargo. The self-unloaders are very flexible in that they can pull up to any dock and deposit their cargo of coal in several hours. However, the vessels without the unloading apparatus are often larger, and faster when operating at ports containing high-speed, large-capacity unloading equipment.

Economy of Operation

Great Lakes freighters, because of their size and intensive use, usually produce as low a transportation rate as any other form of transportation in the world. When aided by restrictive measures such as freight rate zones, their ability to move coal or oil products is enhanced. For example, some coal mined in southeastern Ohio, West Virginia, and Pennsylvania is transported to the southern end of Lake Michigan via Lake Erie ports and lake freighters, instead of by a more direct rail route only half as long.

Environmental Impact

To take advantage of such low-cost transportation, as well as the availability of water, many industrial and electric power plants have been built along the shore of the Great Lakes system. This has intensified environmental pressure in some areas, such as the southern end of Lake Michigan. In general, the lake freighters do not create much of a threat to the environment. There is coal dust blowing about during the loading and unloading operations, and always the threat of a ship sinking. Outside of the port areas, these ships with their covered hatches create beauty and interest as they sail the Great Lakes. Perhaps the enforcement of the use of holding tanks, to be emptied into port sewage systems, would further enhance their acceptance.

Ocean Colliers

In some parts of the world, the ocean collier is operated in direct competition with overland transportation. In the United States, coastal colliers carry coal from Hampton Roads, Virginia, to ports along the northeastern seaboard. Ships are used because, in comparison to railroads, the cost of carrying the coal is lower. In Great Britain, ships also carry coal from the northeast coal fields to markets in southern England, a task which could be accomplished by trains, if it were not for the higher cost.

In other situations, the collier functions to transport coal from one continent to another, with no competition from any other transport media. Since the United States possesses such large quantities of excellent quality coking coal, it exports millions of tons of coal each year, especially to the Common Market countries and Japan. As a result of this trade, Hampton Roads, Virginia, has become the leading exporter of coal in the world.

The environmental impact is much the same as with the lake freighters. There have been few sinkings of colliers over the decades, and dust and spilled coal remain a problem in the port areas. The greater size of the ocean eliminates a concern over sewage discharge from the ship while under way. Restrictions on such discharge should be required while the ship is in port, estuary, or otherwise close to shore.

Pipelines

In the fifties, certain electric power companies began to investigate alternative ways to transport the increasingly larger amounts of coal required at their thermal power plants. This was especially true for those plants which relied solely on rail transport at a time when the railroads were raising their rates in order to meet rising costs of labor and materials. One such utility, Cleveland Electric Illuminating Company, of Ohio, U.S.A., decided on a bold venture: the construction of a pipeline which would carry coal in a slurry mix.

The pipeline was built from the coal-producing area near Cadiz, Ohio, in 1955-56, to CEI's Eastlake Plant, a few miles east of the heart of Cleveland, Ohio. The rail freight rate on coal had been $2.63 per ton, but had risen to $3.47. This forced the utility to try something innovative, as a pipeline. The 108-mile-long pipeline, costing 6,500,000 dollars including the right-of-way, began operation in 1957. It was designed to transport 1,500,000 tons of coal per year, which had been ground to a one-sixteenth-of-an-inch diameter. Mixed with water to form a fifty-fifty slurry mix, the coal was pumped the full distance in thirty-two hours. This proved that pipeline transportation of coal was feasible, dependable, and economical.

On August 17, 1963, this coal pipeline was shut down. The railroads, using some of the bargaining factors made possible by the passage of the Smathers Bill, offered a rate of $1.88 per ton-mile, which was cheaper than the cost via pipeline. Nevertheless, the pipeline served as a model for the next major coal pipeline in this country.

Black Mesa Pipeline

As a result of the previous success of the Ohio pipeline and the expense of providing alternative transportation, the decision was made to build a pipeline to transport coal from the Black Mesa coal mine of northeastern Arizona to the Mohave power plant, located in southern Nevada, just across the Colorado River from Bullhead City, Arizona. This pipeline is designed to carry five million tons of coal a year, through the 273-mile, mountainous route. It began operation in the summer of 1970 and again has proven the applicability of pipelining to coal transportation.

Environmental Impact

Coal pipelines produce the least effect on the landscape of any transportation media. Buried a few feet beneath the surface of the ground, releasing no heat to change the soil characteristics, slurry pipelines can become almost invisible once the natural vegetation has grown back over the area. Pumping stations, pipeline markers, exposed cuts where blasting or ditching was necessary, as well as a straight-line impression across the landscape indicate in some places that this is the route of the pipeline, but the net effect on the environment is minimal.

Conveyor Belts

The continued development in size and capacity of conveyor belts has expanded their use in the transportation of many commodities, including coal. Originally used to move coal at processing plants, or very short distances between mine mouth and power plant, conveyors are today used in partnership with rail and barge, bringing coal across hilly landscapes down to the river's edge or to the railroad track. The construction of such conveyor belts, many miles in length, has helped develop coal deposits which were not otherwise as attractive from an economic standpoint. It might be added that, in strong competition with the Ohio pipeline, an elevated, enclosed conveyor belt was proposed, but this met with strong opposition from the rather densely populated areas it was to traverse. By using the latest technological developments and enclosing the belt, the noise and dust problems can be largely eliminated; but the visual assault on the landscape causes vigorous opposition.

Competition

Transporting coal is an important business in many parts of the world. In some areas, it is possible for more than one form of transport to vie for the business. In the United States, west of the Mississippi River, the train is most important because of the long distances between mine and market. East of the Mississippi, although the railroad is still extremely important, areas do exist in which truck, barge, and lake freighter are able to compete. In some cases, different rates are charged for identical distances between the same coal field and two different markets. A frequent explanation is the presence of another railroad, or inland water barge, able to move the coal as cheaply as the first railroad. Thus, competition enters in, driving down the freight tariff, to the benefit of the purchaser of the coal. Sometimes government intercession is necessary to keep one company from dropping their rates so low as to drive the competition out of business.

Chapter 5

Pumping Petroleum

Since the drilling of the first well in the United States by E. I. Drake in Titusville, Pennsylvania, in 1859, oil and its refined products have supplied an increasingly greater percentage of the total energy consumed by mankind. In 1971, the Free World turned to oil for 54 percent of its energy needs, about twice the percentage of the next most important energy source, coal. Its major use is as fuel for transportation equipment, industrial power plants, and thermal electric power plants. Its advantage is that of containing a large amount of energy per pound when compared with other fuels—about 19,000 BTU's. Consequently, the demand for petroleum continues to increase each year.

Origin and Characteristics

It is generally believed that oil was formed from the remains of very small animals and plants which lived in the shallow seas hundreds of millions of years ago. As these died in great quantities, their remains drifted to the bottom of the seas where they mixed with sand and mud to become marine sediment. Later, this was covered by layers of other sediment which eventually turned into rock. The theory is that, with the passage of time and the movement of the earth's crust, heat and pressure were brought to bear upon the plant and animal remains, and that, combined with their decomposition, formed oil.

Crude oil varies considerably from oil field to oil field. In general, it is composed by weight of 95 percent hydrogen and carbon compounds with a variety of other substances, most notably sulfur. Crude oil containing hydrogen sulfide is referred to as sour petroleum. Crude oil is also differentiated by the composition of its base, which can be asphalt (tar), paraffin (wax), and mixed base. In general, crude petroleum is a dark green or greenish black color, although some oils are pale yellow or red.

Geology

Once the oil has been formed, it must be concentrated into a pool which is large enough to warrant the cost of development. Pools are created by geologic traps. These can be divided into three groups: structural, stratigraphic, and structural-stratigraphic (see figure 5.1). Structural traps are created by the faulting and folding of rock layers, while stratigraphic traps are the result of gradation. More than 75 percent of the world's known oil deposits are in structural traps, mainly because these are easier and cheaper to find when compared with the stratigraphic traps. It is now thought that an increasing amount of oil to be discovered in the future will be in stratigraphic traps.

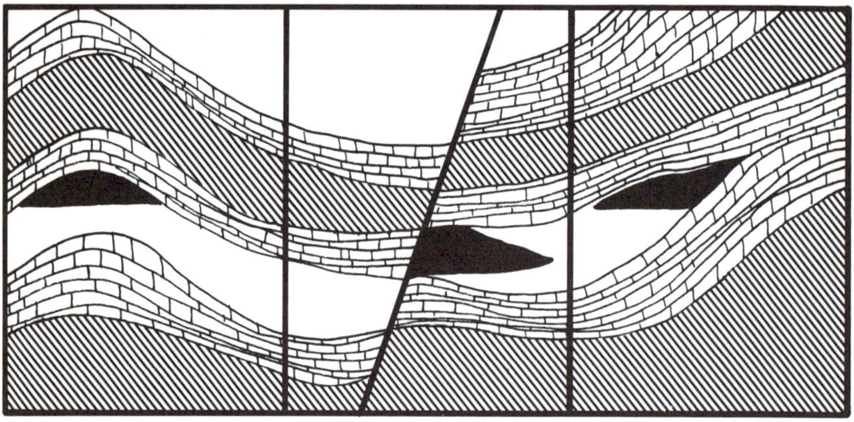

ANTICLINE FAULT STRATIGRAPHIC TRAP

FIGURE 5.1. The three most common oil traps

The major kinds of structural traps are anticlines, and certain conditions associated with salt domes and faults. The first of these is the easiest to discover, although the salt domes which are common to coastal Louisiana and Texas are relatively obvious to the trained geologist. The most significant stratigraphic trap is the reef. These are old, very porous limestone beds—the remains of coal organisms which formerly inhabited shallow, warm seas. Traps are found at varying depths, vary considerably in size, and may or may not contain oil.

Exploration

The exploration for petroleum is a time- and capital-demanding activity. It may easily cost one corporation more than a million dollars a day to support the men and equipment in their search on land or offshore for new oil discoveries. With the emphasis today upon offshore and arctic exploration, costs have gone up rapidly in the last decade. A *wildcat* well is one that is drilled with the hope of discovering a new pool of oil. This is in contrast to a *production* well, which is drilled to increase or somehow facilitate the production of oil from a known pool or field. The risk on production wells is low, whereas wildcat wells are very high. In the United States, only about 10 percent of the wildcats ever find oil, even after tremendous amounts of time and money have been spent on the choice of a likely drilling spot.

Petroleum geologists agree that the easily accessible oil fields have been discovered and that in the future, new discoveries will be found offshore, in areas with difficult access, such as northern Alaska, the western Amazon Basin, or deeper beneath the earth's surface. The deeper the wells are drilled, the greater the expense, and consumers around the world can expect to have to pay more for oil in the coming years.

offshore drilling

Man has been drilling for oil under the seas for the past half century. Every year sees him going to areas having more difficult physical conditions, and deeper water. This has necessitated the rapid development of offshore drilling platforms. These include platforms (which are floated in place and then extend their legs down to a firm base), floating platforms, drill barges, and drill ships. Such progress brings with it increased expenses and greater stress on the environment.

depletion allowance

In the United States, the federal government has long encouraged the search for oil, in part by the use of the depletion allowance. At present, it allows the corporation to base its income tax on net profit minus 24 percent of the value of the oil produced at the wellhead. From the time of its inception until 1969, the rate was set at 27½ percent, but continued cries of favoritism, tax loophole, etc., led to Congress reducing that rate to the present 24 percent. It is interesting to note that this action was taken at just about the same time that the general public was becoming aware of the energy crisis, or petroleum shortage in the United States. The actual utility of this rule in encouraging the search for oil in the United States is open to heated debate.

Drilling

Most of the world's oil wells are less than 20,000 feet deep. The depth of the well, and the condition of the rock through which the well must be drilled, largely determine its cost. Shallow wells sunk in soft rock may only cost in the tens of thousands of dollars while a deep well in hard rock may run 250,000 dollars. Offshore rigs can easily exceed that figure because of the daily expense of operating the specialized equipment involved.

the american experience

In the early days of the American oil industry, most oil wells were relatively shallow, and the derricks made from wood or iron. If one individual took the initiative and drilled a well, striking oil, his neighbor was certain to follow suit—in fact, all his neighbors and perhaps the neighbors of their first neighbors who were successful in striking oil (see figure 5.2). Now, as soon as the first man realized that his neighbors would drill wells, also discover his oil field, and begin to pump out his oil, he drilled a sec-

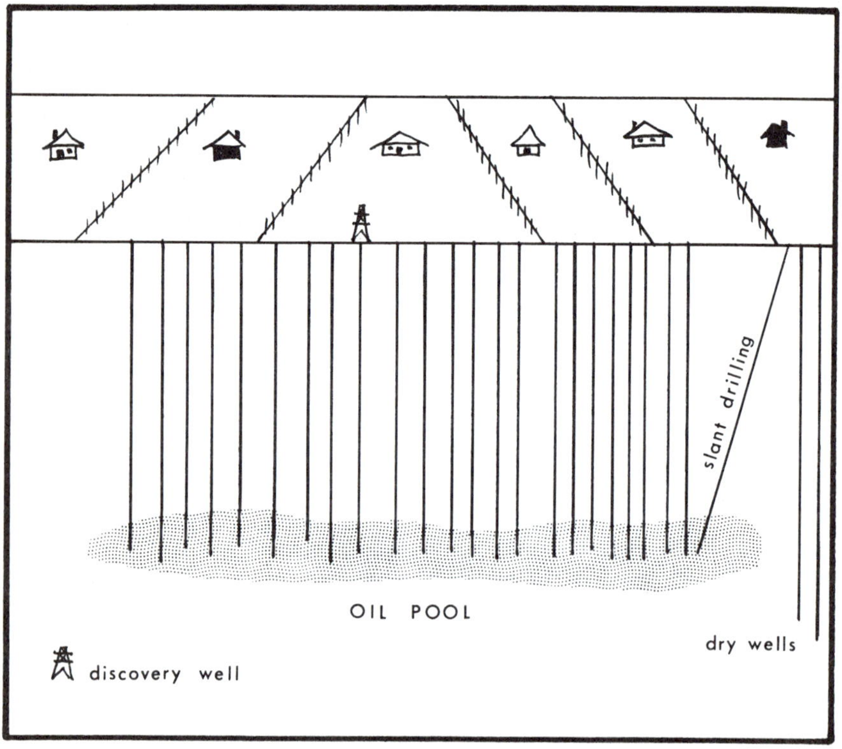

FIGURE 5.2. Unregulated oil well drilling

ond well. And on and on this mad circle of events would continue until oil derrick abutted against oil derrick. The reason for this is that the production of this resource comes under the legal heading of "rule of capture." In other words, just because one man had the initiative to risk drilling for oil does not insure that man the right to the oil beneath his property. If he chooses not to pump in oil out of the ground, his neighbors can not only pump their oil out, but remove all of the oil from beneath his property. If one of the neighbors drilled a dry well, discovering that the oil pool did not extend beneath his property, his next well could prove successful by the practice of *slant drilling.* By drilling at an angle, it was often possible to tap into the pool beneath your neighbor's property. This type of drilling is no longer allowed in this country, if the objective is to obtain oil not legally yours. Thus, the discovery of oil brought about a mad scramble to sink as many wells into the pool as possible in order to insure getting your share, or more, of the oil.

field energy

An oil field is under pressure. The oil is flowing or flowed through a permeable layer of sedimentary rock, trapped above and beneath by impermeable layers of rock. Mixed in with this crude oil is water and natural gas. Since the density of the natural gas is the least, it rises to the top. Since the density of oil is less than water, it tends to float on top of the water. The squeezing of the oil between the layers of natural gas and water represents the field energy. If it is high enough, the wildcat well which punctures it may bring in a gusher, the delight of which movies dramatize because of the roar and spectacle of oil shooting high into the air.

If this field energy is maintained, the oil can be removed from the ground initially by simply opening and closing the system of valves which are installed on the wellhead. These are known as Christmas trees. But once enough natural gas and water have been removed, the field energy declines to the point where pumps are necessary to pump the oil up to the surface. This results in increased production costs and a smaller yield of oil from that field.

unit pool management

To offset the expense and difficulties of unlimited drilling in an oil field, the large oil corporations have shown the way in practicing unit pool management. Once oil is discovered, the appropriate corporations drill production wells to determine the extent of the pool, and the percentage of the oil under each corporation's leased land. Once determined, a few wells on the perimeter are chosen to pump all the oil, and this production is then divided according to the predetermined percentages. In this way, less capital investment is needed to produce the oil, and the development of the field is as if it has one owner—unit pool management. Although more unit pool management is taking place in the United States, its dominant use has been in the great discoveries of oil in the Middle East and North Africa. To dramatize the contrast, in the mid-sixties, the United States wells produced an average of 11.5 barrels per day whereas the wells in Bahrein were averaging 20,000 to 30,000 barrels per day. At the same time, 350 wells in Kufait were producing 32 percent of the Middle East oil production.

A parallel development of unit pool management has been the attempt to manage the field pressure other than by just limiting the number of wells. This involves the reinjection of the water and natural gas back into the pool via nonproduction wells. Work continues along these lines in an attempt to find better ways to remove the remaining oil that is not concentrated sufficiently to be pumped to the surface. In the United States, production quotas per field, per well, have been used, along with limiting the density of oil wells. These acts have combined to preserve somewhat the field energy and to extend the life of the field. The practice of unit pool management coupled with the spatial concentration of oil in the Middle East and North African fields has encouraged the international oil companies to prefer this area for production and further exploration, rather than tackling the more difficult and, hence, more expensive challenge of finding new, and most likely deeper fields in the United States.

Environmental Impact of Oil Produciton

In contrast to the development of resources, such as coal, oil production is a punctiform economic activity. It needs little space and once today's portable drilling rig has been removed, and the well topped by a Christmas tree, which stands only a couple of feet high, there is relatively little visible scar to the landscape. Somewhere oil pipelines must run, connecting the wells with the oil storage tank, and often the water and gas separation tanks. Some oil spillage stains small areas around the pumps and the tanks, but vegetation, either natural or agricultural, can soon cover up most of the scars in a humid region. In dry areas, which naturally support only a cactus or grass, the activity is not as well hidden.

Subsidence

To some, the mere presence of the Christmas trees, tanks, pipelines, etc., of a modern, well-developed oil field are still offensive. But perhaps the most serious effect on the environment comes from subsidence. The removal of the oil, along with the water and natural gas, reduces the pressure which in part has been holding the overlying rock layers in place. The result is subsidence. A classic example of this phenomena is found in the Long Beach-Wilmington region of southern California. The sinking of the ground took on an elliptical shape, which covered twenty-two square miles and sank as much as twenty-six feet. This resulted in damage and destruction to buildings, highways, pipelines and the like, as well as flooding of areas which formerly had been above sea level. Much research and experimentation is now going on to find ways of maintaining sufficient pressure during and after oil extraction, to prevent subsidence from taking place.

World Production

The first commercial oil well was drilled in 1857, near the city of Ploesti, Romania. The first year's production of 2,000 barrels represented the world's entire production. Two years later, near Titusville, Pennsylvania, the first commercial oil well in the United States was drilled. Production from this field was 500,000 barrels in 1860, and this represented the domination of production which was to continue for many decades. As late at 1925, the United States produced 71 percent of the world total; as recently as 1955, 44 percent; and in 1969, 22 percent. The trend suggests this figure will slide into the teens within the decade. This decline has taken place, not because of any reduction of production in the United States, but is, rather, due to the discovery of major oil fields around the world, and their subsequent production to meet the expanding world market. It should be noted that the leading seven nations accounted for almost 75 percent of the total production in 1969 (table 5.1).

United States

With the success of that first commercial oil well in Pennsylvania, exploration continued with the successful drilling of other wells in that State as well as New York, West Virginia, and Ohio. By 1900, Ohio had become the leading State for several years before losing out to Oklahoma. By 1910, California was leading the states, with Oklahoma in second place, Illinois in third, and Ohio dropping back to fifth, due in part to a better than 50 percent decline in production. By 1920, the present pattern began to emerge with Oklahoma first, California second, and Texas third. For more than three decades, Texas has been the leader, with California, Oklahoma, and finally Louisiana holding down the number two position.

The ten leading states produce almost 95 percent of the total American production (table 5.2). Of these states, only Texas and Louisiana have

TABLE 5.1

The World's Ten Leading Oil Producing Nations, 1971

(thousands of barrels)

Rank	Country	Production
1.	United States	3,453,914
2.	U.S.S.R.	2,778,300
3.	Saudi Arabia	1,741,149
4.	Iran	1,661,901
5.	Venezuela	1,295,406
6.	Kuwait	1,167,329
7.	Libya	1,007,687
8.	Iraq	624,312
9.	Nigeria	558,375
10.	Canada	495,743
	World Total	17,634,608

SOURCE: The World Almanac, 1973 edition; published by Newspaper Enterprise Association.

TABLE 5.2

The Ten Leading Oil Producing States, 1970

(millions of barrels)

State	Production	Average Production Per Well, Per Day	Proved Reserves
1. Texas	1,142	19.0	13,023
2. Louisiana	907	86.7	5,399
3. California	372	25.0	3,706
4. Oklahoma	225	7.7	1,405
5. Wyoming	160	48.1	997
6. New Mexico	128	20.4	657
7. Kansas	85	5.3	502
8. Alaska	84	1,190.1	10,116
9. Mississippi	65	63.2	342
10. Illinois	44	4.5	209
U.S. Total	3,517	U.S. Av. 18.0	U.S. Total 38,063

SOURCE: *Statistical Abstract of the U.S.*, (Washington, D.C.: U.S. Department of Commerce, 1972).

been able to increase their production in recent years, but Alaska can be added to that list once the transportation problem is resolved for its North Slope oil. The list of the leading oil fields in the United States indicates the strength of the leading producer states (table 5.3).

U.S.S.R.

The U.S.S.R. has a long history of oil production, having led the world at the turn of the century. But the rapid growth of America's oil industry

TABLE 5.3

Production and Reserves of the Ten Leading Oil Fields in the United States, 1971

(thousands of barrels)

Name	State	Production	Estimated Reserves
1. Wilmington	California	72,859	900,500
2. East Texas	Texas	71,139	1,986,020
3. Kelly-Snyder	Texas	52,487	619,232
4. Wasson	Texas	51,210	848,790
5. McArthur River	Alaska	40,683	190,935
6. Sho-Vel-Tum	Oklahoma	36,500	243,500
7. Slaughter	Texas	35,515	284,485
8. Midway-Sunset	California	33,583	526,417
9. Caillou Island	Louisiana	31,828	297,333
10. Timbalier Bay	Louisiana	30,988	319,519

SOURCE: *Minerals Yearbook, Vol. I, Metals, Minerals, and Fuels,* (Washington, D.C.: U.S. Department of the Interior, 1971).

soon exceeded the Russian output while a series of events hindered Russian expansion. Since 1953, much emphasis has been placed on this sector of the economy, and production has increased at a rate of more than 20 percent each year. During the period, a major shift of producing areas has taken place. Whereas in 1940, 87 percent of total production and 80 percent of reserves were in the Caucasus region, by the mid-sixties, more than 70 percent of total production and 81 percent of reserves were in the newer Volga-Urals oil fields. In the Caucasus, the Baku fields led the nation in production until 1954. Then these fields experienced a decline. Recently they have experienced a resurgence back to pre-World War II levels, with the possibility of some expansion due to new discoveries on the fringe of the fields. But the Volga-Urals or "Second Baku" is the growth area with three political areas, each surpassing the Baku fields' production. Of these, the Tatar A.S.S.R. is the leading producer. Time and further exploration will reveal if the newly discovered oil fields in the Ob River valley will eventually surpass that of the Volga-Urals region.

Venezuela

The fifth leading oil producer, Venezuela, pumps about 80 percent of her oil from the oldest and largest field, located in the Lake Maracaibo area of western Venezuela. This oil, mainly found in stratigraphic traps, is mostly concentrated on the eastern side of the lake, both under water and land. As a result, in drilling for oil beneath the water during the 1920s, new techniques were tried and perfected which have since been used to

TABLE 5.4

World's Largest Oil Fields by Ultimate Recoverable Oil, 1969

(billions of barrels)

Rank	Field and Country	Discovered	Ultimate Recoverable Oil
1.	Greater Burgan, Kuwait	1938	62
2.	Ghawar, Saudi Arabia	1948	45
3.	Bolivar Coastal Field, Venezuela	1922	30
4.	Safaniya-Kafji, Saudi Arabia Neutral Zone	1951	25
5.	Kirkuk, Iraq	1927	15
6.	Rumaila, Iraq	1953	13.6
7.	Agha Jari, Iran	1938	9.5
8.	Abqaiq, Saudi Arabia	1941	9
9.	Gach Saran, Iran	1928	8
10.	Sarir, Libya	1961	8
11.	Raudhatain, Kuwait	1955	7.7
12.	Minas, Sumatra	1944	7
13.	Manifa, Saudi Arabia	1957	6
14.	Marun, Iran	1957	6
15.	Ahwaz, Iran	1959	6
16.	East Texas, U.S.A.	1930	6
17.	Bibi Hakimch, Iran	1962	4.5
18.	Sabriva, Kuwait	1958	4
19.	Murban Bu Hasa, Abu Dhabi	1962	3
20.	Faris, Iran	1964	3
21.	Qatif, Saudi Arabia	1945	3
22.	Lama, Venezuela	1957	2.7
23.	Hassi-Messaoud, Algeria	1956	2.7
24.	Wilmington, California, U.S.A.	1935	2.6
25.	Khursaniya, Saudi Arabia	1956	2.5
26.	Wafra, Neutral Zone	1953	2.5
27.	Abu Sa'fah, Saudi Arabia	1963	2.5
28.	Dukhan, Qatar	1940	2.4
29.	Zaltan, Libya	1959	2.2
30.	Poza Rica, Mexico	1930	2
31.	Minagish, Kuwait	1959	2
32.	Gialo, Libya	1961	2
33.	Duri, Indonesia	1941	2
34.	Murban Bab, Abu Dhabi	1954	2
35.	Umm, Shaif, Abu Dhabi	1958	2

SOURCE: *Oil and Gas Journal*, January 13, 1969, pp. 46-47.

drill for oil in the shallow waters of many continents. The rest of the oil is obtained from newer but smaller fields, located north of the Orinoco River in eastern Venezuela.

About 80 percent of the oil production is in the control of three large corporations: Creole Petroleum Corporation (a subsidiary of Standard Oil of New Jersey), with 40 percent; Royal Dutch Shell, with 25 percent; and Mene Grande Oil Company (owned by Standard Oil of New Jersey, Royal Dutch Shell and Gulf Oil Company), with 15 percent. About 75 percent of the crude oil is exported from Venezuela, with 25 percent going to the United States refineries and 50 percent going to the refineries on the islands of Aruba and Curaçao. The refined products from the island refineries go to the Latin American and European markets. Venezuela is building towards the day when she will control the whole industry, drilling, producing, and refining her oil.

The Arab Nations

The production of oil in the Middle East began about the turn of the twentieth century. Although much greater potential was suspected of the Middle East countries, it was not until the post-World War II era that large-scale development proved this (table 5.4). The major fields were discovered in two areas of concentrations; the first and most important were located in areas to the north, west, and south of the Persian Gulf, while the second concentration was mainly in northeastern Iraq. During the sixties, extensive fields were discovered in North Africa, in particular, Zaltan in Libya, and the Polignac Basin and Hassi-Messaoud in Algeria. As a result, oil production has become an important function in almost every Arab nation. In 1971, Saudi Arabia had increased its production so as to rank as the leader among the Arab nations. Iran, Kuwait, Libia, and Iraq had production totals that were not far behind. It was even suggested that Libya with its meteoric rise to seventh place on the top ten oil-producing nations list within about a decade, might soon become the leading Arab nation. However, the government decided to impose pumping restrictions and thus the production rate has been held at about the 1971 levels through the present.

One characteristic of oil production in the Arab nations has been the dominance of control in the hands of foreign corporations. Traditionally, individual corporations would bargain with the nations, one at a time, in order to determine the royalty to be paid to the government of each producing nation. The rest of the business was left to the corporate management, and many Arabs felt they received less than their fair share of the profit. As a result, the Organization of Petroleum Exporting Countries (OPEC) was founded in 1960 by Iran, Venezuela, and Saudi Arabia.

Although Venezuela is not an Arab nation, it was experiencing the same treatment in bargaining for royalties with the large corporations. These three were soon joined by the other major producing nations in the Arab World, creating an organization which has been most successful in obtaining higher prices for their oil. For example, between 1970 and April 1973, the price paid to the Arab governments by the oil companies has increased 72 percent per barrel.

Arab-Israeli Conflict

Not only are the Arab nations gaining greater economic returns from their petroleum resources through a united front presented by the OPEC, but further steps are being taken by the individual nations to nationalize part or all of the petroleum industry within their political territory. Thus lower profits and the threat of nationalization of their assets concern the managements of the various petroleum corporations operating in the Middle East and North Africa. Yet another problem lurks in the background; the unrest and uncertainty associated with the State of Israel which was imposed by outsiders on the Arab World. The establishment of Israel (in 1948) brought about such problems as the displacement of Arabs, who were settled in refugee camps. This problem still remains unresolved.

At present, the United States is a major supporter of the freedom and existence of Israel, and this attitude irks the Arab world. With the present concern over sufficient supplies of oil to keep not only the United States but Western Europe properly fueled, it will be interesting to watch the maneuvers undertaken in the field of international diplomacy to unravel this situation.

Arab Wealth

The Arab States are becoming increasingly important, not just because of their ability to produce the oil needed by other parts of the world, but also because of the increasing concentration of capital earned by means of the sale of the oil. In 1968, the income of all the oil-producing countries of the Middle East and North Africa, all of which are Arab except Iran, was $4.4 billion. In 1972 it was more than $10 billion, and by 1980 might be as much as $40 billion. Since about 60 percent of the world's known reserves are possessed by these countries the future looks very promising. In the meantime, speculation runs rampant as to how this wealth will be spent, and what effect it will have on world economies and international relations.

North Sea Oil Discoveries

One of the most exciting developments of recent years has been the discovery of oil and natural gas beneath the North Sea. The first major clue that such resources might in fact exist was the discovery of the Groningen gas field in northeast Netherlands. This triggered interest in off-shore exploration, and created a need to establish territorial rights to the continental shelf beneath the North Sea waters.

In response to this need, those nations bordering the North Sea adopted a new international convention under the auspices of the United Nations. This included the extension of the width of the territorial seas claimed by each nation from 3 miles to 6 or 9 miles and the division of the North Sea into sectors of different sizes, according to the length of the seacoast and the median lines between the nations. As a result, Great Britain, Norway, Denmark, Sweden, West Germany, Belgium, and France share in varying degrees the ownership of the bottom of the ocean floor.

The first payoff was the discovery of natural gas off the Norfolk coast of southern England. After continued efforts, a major oil field was discovered in December 1969. This field, named Ekofisk, is located within the territory assigned to Norway, and should enable that country to change its position as a net importer of oil to that of a net exporter. The field is considered to contain 1 billion barrels, with some estimates suggesting as much as 3 billion. It is now producing 11,000 barrels of oil per day, but once the field is fully developed, that figure will increase to a planned capacity of 300,000 bopd.

In October 1970 the next major discovery was made. Located in the British sector, it is known as the Forties Field, and is estimated to contain 2 billion barrels of oil. In July of 1971, the Brent Field was discovered. It is also located in the British sector and contains 1 billion barrels of oil. Discoveries such as these spur the oil companies on to further exploration. The cost of such efforts is enormous, and these fields will not be as profitable as those in the Middle East or North Africa. For example, the cost of a wildcat well ranges from $750,000 in the southern part of the North Sea to 5 million dollars in the northern part. The depth of the water requires new types of drilling rigs, far more sophisticated than the now famous Texas Towers that were used in the shallower Gulf of Mexico. Operating a semisubmersible drilling rig now costs $45,000-$50,000 per day.

The demand for such drilling rigs, and their support facilities, has suddenly increased the tempo of activity at shipyards and ports in the North Sea area. Drilling towers which were formerly made in Texas and towed to the area are being built in the shipyards of Europe, in particular northern England and Scotland. Many relatively quiet North Sea ports, such as Aberdeen, have come alive with manufacturing and service activities necessary for the continued exploration and development of the North Sea oil fields.

The 42 billion barrels of oil beneath the North Sea (current estimate) will find ready markets in the European nations. It will reduce, but not eliminate Europe's dependence on Middle Eastern and North African oil fields.

Chapter 6

Moving Petroleum to the Market

To an even greater extent than coal, oil is found in large quantities in just a few parts of the world. Many of these outstanding oil deposits are located in underdeveloped nations, separated from the major oil consumers by oceans or extreme land routes. Thus it is natural to understand why the oil tanker is so important today, yet this has not always been so.

In the earliest days of the oil industry in the United States, the oil discovered in western Pennsylvania was distant from any major market. The obvious answer was to use the existing railroads to carry the oil east to the coastal markets. It was, therefore, necessary to transport the oil to the railhead, a few miles away. This was accomplished by loading the oil into wooden barrels, and then carrying them by horse and wagon to the railhead. Since so much oil leaked out from the barrels and the teamsters charged a fairly high rate, an innovative man came along and offered to do the job for much less. His solution, which was adopted, was the construction of a wooden pipeline, made from hollowed-out logs. Even though oil leaked from the poor seal at the joints, it still could do the job for less. And that's the way competition has been working ever since, to get oil from the fields to the markets.

Trucks

The early teamsters, with their horses and wagons, eventually gave way to motorized wagons, or trucks. Today the truck plays an important role in the movement of oil, but its particular role has changed. Its main function is the distribution of refined products to the retailers and consumers: to the gas stations, and to the fuel oil tanks in homes and other buildings. Only in rare cases or emergencies are trucks used to transport crude oil.

Railroad Tank Cars

The railroads were a vital link in the movement of by-products from the refinery to the markets. As long as the scale of production and consumption remained rather small, such transportation remained competitive. In the United States, the concentration of refining capacity at certain centers and the large-scale production of refined products made it more economical to move such volume by pipeline. Today, the railroad tank car is important in transporting refined products to areas not served by pipeline or tank truck. At the present time, the rapid addition of oil-burning peaking units by electric utilities and the shortage of natural gas to continuously fuel the gas-burning thermal power plants have created a demand for oil, which either does not justify the cost of building a pipeline to serve the plant or which has not given the lead time necessary for the completion of such a project.

In other parts of the world, where railroads exist, their use to carry refined products is still economical today, mainly because of the scale of operation. The developed nations of the world will eventually be served by product pipelines, which will be supplemented by tank truck deliveries to the retailers and consumers.

Pipelines

Pipeline transportation of oil—crude and refined—is the most important overland method. This is due to the economics of pipeline operation. Once in the ground, pipelines are simply more efficient, less labor demanding, and require less maintenance than any of the alternatives. To justify their construction, sufficient oil or refined products must be available from oil field or refinery complex to keep the pipeline operating for at least twenty, and usually thirty, years. The cost to transport crude and refined oil can be the lowest of all media, when the pipeline is used to capacity throughout the year.

A good example of pipeline economics is represented by the product pipeline connecting the refineries in the Los Angeles metropolitan area with Phoenix, Arizona, in the United States. The small market simply does not justify the construction of a modern refinery in the Phoenix area, nor does the market share of any one company justify the construction of a pipeline to be used by only one company. Instead a pipeline was built by Southern Pacific Pipeline, from Norwalk, California, to Phoenix, a distance of 418 miles. The refineries of the different oil companies deliver their by-products to Norwalk, where it is then sent through to the appropriate tank-storage farms in Phoenix. The by-products are sent through in batches—a batch of regular gasoline for each of the companies, to be followed by batches of high-test gasoline, low-lead gasoline, fuel oil, etc. By grouping the same by-products in consecutive batches, the quality of these products is better maintained. It should be added that Phoenix is also served by a similar pipeline, 427 miles long, from the refineries at El Paso, Texas.

Throughout the world, in the developed nations of Europe or in the underdeveloped nations of South America, new pipelines are being built to move crude oil out of the interior to world shipping routes, or into the interior to refineries. In many parts of the world, pipelines are being built to move the refined products from coastal refineries to inland markets, replacing the more traditional rail, truck, or barge transport.

environmental impact

As with coal pipelines, crude and product pipelines inflict a minimum disturbance on the environment. There is the danger of leakage, but this can be kept to a minimum by apparatus which determines if the pipeline has ruptured from some natural or man-made action. It then causes the pumps to be shut off, stopping the further flow of oil. A major difference between coal and oil pipelines is the need for the oil to be heated to improve its viscosity, enabling it to flow through the pipeline easily. In applications in the middle and low latitudes, this presents no real problem. But in the high latitudes, there is a different set of problems.

The Alaska Pipeline

Perhaps the most controversial pipeline in the world today is the Alaska pipeline. This four-foot-diameter pipeline would carry two million tons of oil per day at a temperature of 145° F, for 789 miles across Alaska, from Prudhoe Bay on the Arctic Ocean to Valdez, on the Pacific Ocean. Perhaps the most disturbing feature of this pipeline is the temperature of the oil. It is argued that such heat will melt the permafrost, causing the pipe to sag to the point of rupturing, as well as causing mudslides downslope along the route. Furthermore, the construction of this pipeline above ground cuts across the migratory routes of major caribou herds, crosses 350 rivers and streams, and lies atop or near a couple of earthquake faults. The reforestation of the pipeline route will require a long time, due to the slow growth rate of the tundra and taiga vegetation of this region, and the opponents suggest that the damage done will further delay or perhaps prohibit its completion.

An alternative route has been suggested that is about three times longer, or 2500 miles. It would be built through the Mackenzie River Valley to Chicago, paralleling an already planned gas pipeline. Such a route would eliminate the threat of oil spillage at Valdez, where the crude oil would be pumped aboard tankers for the trip to West Coast cities, but it would delay the project by two to three years. Both sides have much at stake in the final outcome of this battle over the construction of the trans-Alaska pipeline, for it will throw light on the prospects of future oil-field developments in the North American Arctic.

Inland Barges

When inland barges transport petroleum, they are usually carrying refined products. It is economical to use barges where refineries or product pipelines have access to a navigable waterway. Such barges can then be pushed or pulled to various markets where they can be left during the period of unloading. These are often used on a seasonal basis because of the problems associated with ice or low water levels.

As an example, by-products from the Bayonne, New Jersey, refineries are transported by barge up the Hudson River and the New York State Barge Canal. During the spring, summer, and fall, these barges can be seen gliding along river or canal waters, or tied up at tank farms and depots located near all the large cities along the canal right-of-way. During the winter, ice usually accumulates to the point of making the trip impossible, and the markets in this area must then rely on tank trucks for their deliveries of refined products, resulting in a higher cost.

Although the cost of transporting petroleum products by barge is very inexpensive, the greater capacity and ease of using pipelines has often resulted in a switch-over to the latter, when the volume increases to an appropriate level. Barges still serve important routes, as well as areas of smaller markets, in many parts of the world.

environmental impact

Barges offer a certain threat to the environment by being susceptible to small spills during loading and unloading operations. Their greatest damaging effect would occur if the barges were grounded, damaged, or sunk in such a manner as to cause the petroleum products to be dumped into the stream. The oily substances, by joining the waters of a river in particular, would be quickly distributed downstream, making clean-up even more monumental than if the spillage had taken place on lake or canal waters with little or no current. The performance of barge transportation throughout the world has been very good, although small spills, noise, and odor are associated with the operation of the tugboats needed to move dumb barges, or the power plants contained in the self-propelled barges. Barges can transport petroleum products, then, with relatively little impact on the physical environment.

Ocean Tankers

Man has long used ships to transport petroleum. At first, when quantities were small, it was carried in wooden barrels, but as the volume increased, it became necessary to use larger, more economical containers. Thus evolved the first true tanker, which had one large hold and could easily load and unload the oil. Experience soon proved that an undivided hold allowed the oil to slosh back and forth, raising havoc with control of the vessel. Thus tanker ships had their bathtub-like holds divided into several small holds, essentially eliminating the navigation problem caused by uncontrolled sloshing of the liquid aboard.

The ocean tanker has undergone extensive changes during the past three decades. The standard tanker of the Second World War was the T2, a ship of 16,600 dead weight tons, which could hold about 135,000 barrels of oil. The demand for oil, brought on by the Korean War and the general economic expansion of the fifties, encouraged the building of larger tankers. By 1959, the Japanese had surpassed the 100,000-dwt mark with the launching of the Universe Apollo, rated at 104,520 dwt.

The closure of the Suez Canal was a tremendous force in stimulating both the number and size of these supertankers. Such ships make possible the transportation of oil from the Persian Gulf to European ports, via the Cape of Good Hope, for a lower cost than would be incurred using the smaller tankers and the Suez Canal route. Presently, Gulf Oil Corporation is leasing six 326,000-dwt tankers to transport oil from the Persian Gulf to a tank farm in southwest Ireland.

Superports

A tanker rated at 160,000-dwt or more is known as a VLCC, or very large crude carrier. One problem inherent with using VLCC's is the sudden elimination of most ports, usually because of a lack of sufficient storage facilities or insufficient water depth. To offset this problem, supertanker ports are being built to accomodate these ships. Bantry Bay in Ireland is an example of a tank farm development, while at Milford Haven, Wales, a refinery is involved as well. To maintain her position as the refining center of Europe, the Port of Rotterdam has been expanded seaward, where the huge Delta Project is being completed. This includes a new port known as Maasvlakte which has the capability of handling 500,000-dwt tankers. A more recent phenomena is opposition against such ports by various environmental groups. The State of Delaware has already passed a bill which would prevent the construction of any superports along her shore, while citizen groups are now pressing for similar action in Maine. Pressure will continue to be exerted to allow for the construction of at least a couple of superports on the East Coast of the United States, to enable the petroleum industry to supply the market with the products demanded.

World Tanker Fleet

Statistics released for the period ending December 31, 1971, show that the world tanker fleet included 3,679 ships, totalling 183,138,000 dwt. Liberia retained its leadership, with 850 ships flying her flag and accounted for 26.2 percent of the total dead-weight tonnage. Three nations were essentially tied for second place—Norway, United Kingdom, and Japan—each with about 12 percent of the total tonnage, but the number of ships varied, due to their different sizes. The United Kingdom had 442 tankers, Norway had 391, and Japan, only 286 (but, obviously, her fleet included many of the newer, and consequently larger-sized, tankers). The world fleet averages just under 50,000 dead-weight tons.

As of early 1973, the largest tanker afloat is the 483,644-dwt Blobtik Tokyo, built in Japan and owned by a British company. However, a contract has been signed for the construction of a 500,000-dwt tanker to be delivered to a Greek shipowner at the end of 1976. A preliminary contract has been signed between a Japanese shipbuilder and a British company for the construction of a 700,000-dwt vessel; and it appears that the one-million-dwt class will be reached within a decade.

sea-lanes

The major sea-lanes used by the oil tankers are as follows: Venezuela to the United States, U.S. Gulf Coast to the East Coast of the United States, Persian Gulf to Japan, Persian Gulf to Europe via (a) the Suez Canal, when open, (b) the Cape of Good Hope, or (c) the Levant to Europe. The Suez Canal has been closed since the summer of 1967. Until its closure, tankers up to the 60,000-dwt class, with a draft up to thirty-six feet, were using it. Once the canal is reopened to traffic, it should be deepened to allow ships of at least 100,000 dwt to pass through. In the meantime, Egypt is losing millions of dollars each year as the needed oil is transported by supertankers around Africa to Western Europe. The other alternative, which is sharply limited by capacity and political conditions, uses pipelines running from oil fields in Saudi Arabia and Iraq to ports in Lebanon and Syria. From thence, it is transported to Europe by tanker. It would appear to be advantageous to most peoples to see the Suez Canal opened for traffic as soon as possible.

environmental impact

Of greatest concern to environmentalists around the world is the threat posed by the sinking or running aground of a tanker. The spreading oil slick which envelops all sorts of living things brings death and deterioration wherever wind and current carries it. The larger the tanker, the greater the potential damage. A basic argument is whether it is better to transport petroleum in many small tankers or in fewer supertankers.

Since increasing amounts of petroleum are being shipped and longer distances are involved, especially with the closure of the Suez Canal, economics favor the use of larger and larger supertankers. The use of more VLCC's demands the construction of ports capable of serving these ships. They may be onshore or offshore facilities but the environmental threat still persists. Many people, therefore, oppose the construction of supertankers and superports because of the preference for dealing with smaller oil spills at existing port facilities, rather than chancing the damage that could result from an accident with a supertanker or VLCC at a new superport, which most likely will have been built at a nonindustrialized site.

Those who oppose increased shipments of petroleum offer no real alternative that is acceptable to the general world population. It should be added that oil spillages do not introduce a permanent pollutant to an area. In general, waterfowl suffers the most from an oil spill, with varying damages to other marine life. The larger, mobile forms of marine life simply flee the affected area and return months later when conditions are acceptable. Other forms of sea life adapt to the presence of oil until nature has had the time to break it down. Certainly, we must take all the steps possible to prevent spillage from tankers or port facilities, but the evidence suggests that we need not eliminate the transportation of oil.

Transport Competition

Competition and technological improvements often force a change or produce an alternative way of accomplishing a given task. For example, the Abqaiq-to-Sidon pipeline is 1,068 miles long, and was originally built to compete with tankers moving through the Suez Canal. Today its thirty-one to thirty-two-inch-diameter pipe, which was first placed in operation in 1950, is competing with the supertankers sailing around Africa between the Persian Gulf and European ports. The Marseille, France-to-Karlsruhe, West Germany, pipeline, 475 miles of thirty-four-inch pipe, was built to compete with tankers carrying crude oil to Rotterdam, where it was transferred to barges for the final leg of the journey up the Rhine to the refinery at Karlsruhe.

In Canada, the increasing size of tankers, and the time involved in the slow journey up the St. Lawrence River to the refineries at Montreal, prompted the construction of a pipeline from Portland, Maine, to Montreal. This reduced the turnaround time for the tankers, and made possible the use of larger, more economical tankers with deeper drafts, something not possible with the present St. Lawrence River channel. In the United States, the largest pipeline, the forty-inch Capline, extends 640 miles from St. James, Louisiana, to Pakota, Illinois. This pipeline transports 350,000 barrels of crude oil a day, taking away the job that otherwise would be accomplished by barges or trains.

a political pipeline

The construction of the Druzhba or "Friendship" pipeline was not solely for the movement of oil between nations; it was also expected to create a dependency on Soviet oil that would tie together the economies of the U.S.S.R. and the four Iron Curtain countries. This 2500-mile pipeline connects the Volga-Urals oil fields with refineries located in Poland, East Germany, Czechoslovakia, and Hungary. It was completed in 1964 and transports almost all the oil exported from the U.S.S.R. to these nations. It has proven to be an economic success in the movement of oil, but its success in a political sense is difficult to evaluate.

Relative Importance of Transport Media

In the United States, about 75 percent of all petroleum products is transported by pipeline, with the remainder moving by watercraft, railroad tank car, and tank truck. On a worldwide basis, ocean tankers are the most important, followed by pipelines. In third place are inland waterway barges, with three river systems being most important: the Mississippi, the Volga, and the Rhine. Only about 5 percent of the total is moved by rail or truck. The great challenge facing those engaging in the petroleum industry is how to transport the increasingly greater amounts of petroleum to the market in a way that will be both economical and harmless to the environment.

Chapter 7

Manufactured and Natural Gas

History

The gas utility industry began in the late eighteenth and early nineteenth centuries. The year 1813 witnessed the formation of the London and Westminster Gas Company, which owned three manufacturing stations and fifteen miles of main, and supplied the gas for the lights that lit Westminster Bridge. Meanwhile, in the United States, after various demonstrations of the versatility of the use of gas, Baltimore became the first city to adopt a commercial gas system and allow the lighting of her streets via a network of pipes laid down beneath the streets. This was in the year 1816. Since that decade, the industry has been established throughout the world with Europe and the United States having the most extensive and densest pipeline network. Most of this worldwide growth was based on manufactured gas.

Manufactured Gas

production

Manufactured gas is produced by heating a high-volatile coal in a metal retort. The gas released from the coal is then cooled and purified before being sent to a storage vessel known as a gasometer or gasholder. There are two types of gasometers which use the weight of a single lid, or inverted iron can, to keep the gas under pressure. The weight of this top takes the place of gravity in a typical water system with its water tower located on a tower or hilltop. Just as water flows out of the kitchen tap when the handle is turned on, so gas flows from the burner when the gas stove is turned on, the gas being forced through the gas lines by the weight (pressure) of the gasometer lids or covers.

The spatial pattern of these early gasworks reflected the availability of coal supplies via railroad or waterway transportation. Many early gasworks served only one town, although in some cases where distance was not great, one gasworks would provide service to other towns, usually those not having good transportation access for the delivery of coal. Gasworks then were located close to the railroad tracks, or on a river, canal, or tidewater site. They were market-oriented.

Manufactured gas varies according to the characteristics of the coal being used and the exact process being employed. Some examples of manufactured gas include coal gases, coke oven gases, producer gas, blast furnace gas, blue (water) gas, carbureted water gas. The BTU content, or heat value, of these gases varies widely. Most manufactured gas used in the homes for so many decades was very dangerous. Gas concentrated in a home or building from a leak could trigger a violent explosion, or the same leak could lead to death by gasification. How many movies of the 1930s and 1940s contained suspense, violence, and tragedy, using manufactured gas as a prop! Its downfall came about, not because of its dangerous characteristics, but because of the eventual availability of natural gas with its advantages.

Natural Gas

characteristics

Natural gas is found alone, or in association with petroleum. It is a colorless, odorless, nontoxic gas which has a calorific value about twice that of coal gas. In order to maximize production, the gas should be withdrawn at a steady rate, but in contrast to oil, gas presents storage problems. Since the majority of wells producing gas are oil wells, large quantities of natural gas are produced according to the needs of the petroleum industry. In the early years of the industry, this gas was a waste product, which was burned up at pit or tower flares. With the construction of pipelines beginning in the twenties, it became a waste product with a market ready to buy it at low prices. Since so much of the expense of exploration was covered by the oil production, the price for natural gas from the beginning has always been very low. The ease and cleanliness of its use, coupled with its low price, made it easy for natural gas to displace oil and coal in most industrial, commercial, and domestic markets when available in sufficient quantities. Since it is free of sulfur, it can be used for direct process heating in certain industries without affecting the product in any way.

development

The drilling for natural gas employs the same basic principles used in the petroleum industry. The well can be perpendicular, or, under certain conditions, directional (slant) drilling is permissible. For example, when trying to tap some offshore gas deposits, directions drilling may greatly reduce development costs, as well as allow for simpler production procedures. Field energy, as such, is best maintained by limiting the number of wells, and producing the gas at a constant rate.

environmental impact

In contrast to oil fields, natural gas leaves no residue around the production area, although the field is marked by pipelines, Christmas trees, and sometimes processing equipment. The area, however, can return to productive uses, such as grazing or crop-growing.

Of more serious import is land subsidence. The removal of the natural gas, like petroleum, reduces the pressure contained in the porous rock layers. As a result, those areas with weakly supported overlying rock layers will tend to sag as the gas is pumped out, causing subsidence. A classic example of this is the Po River Valley in Italy. The long history of gas production, coupled with the recent weak layers of sedimentary rock typical of this area, has resulted in serious subsidence. Some scholars suggest that this is partly responsible for the continued sinking of cities in the area, Venice in particular.

Gas Production in the United States

The first natural gas well dug in the United States was completed by William Aaron Hart, who investigated and pursued the problem of bubbling water encountered by a man who was digging a water well. In 1821, on his third try, Mr. Hart encountered natural gas, at a depth of seventeen feet. After extending the well to twenty-seven feet, he capped the well crudely and, by means of a wooden pipeline, transported the gas to a gasometer. Lead pipes later replaced the wooden ones which distributed the gas to several buildings in Fredonia, New York, including the Abell House which was located on the stagecoach route between New York City and Buffalo. The inn became famous not only for its brilliant lighting which was far brighter than lamps burning whale oil, but also because this was the first recorded commercial use of natural gas for cooking. Thereafter, many establishments took advantage of natural gas discoveries. For example, near Westfield, New York, on the shores of Lake Erie, a half-mile long wooden pipeline was laid to carry natural gas from a new discovery to the lighthouse. This gas field supplied fuel to be burned in the 144 burners in the beacon for more than twenty-five years.

An obvious advantage to the United States was the presence of huge natural gas reserves. These were first discovered in the early part of the twentieth century, usually as a result of drilling for oil. One such discovery was the Hugoton Field in 1920, which even today remains second only to the Groningen Field in the Netherlands as the largest dry gas reservoir in the world. As the reserves of natural gas expanded with new discoveries, pipelines were built to carry the gas to new and more distant markets. As late as 1944, manufactured gas still dominated production in the United States.

After the end of the Second World War, the construction of more than 100,000 miles of pipeline brought natural gas to all parts of the nation. Today little manufactured gas is produced. The advantages of using natural gas have, in the last several years (in light of the attempt to reduce pollution, especially by industry) encouraged its consumption to the point where the United States can no longer meet the demand with indigenous supplies.

In the United States two-thirds of the total natural gas production is from those fields located in Texas and Oklahoma, with the six leading states producing more than 90 percent (table 7.1). Ninety percent of this gas is sold for distribution to consumers, with the remaining 10 percent being flared, or pumped back into oil fields to help maintain field energy. Of the total natural gas reserves, two-thirds are classified as being nonassociated, that is, not associated with oil deposits. The future prospects for natural gas production seem to be dependent upon the continued limitations on imported natural gas and on whether or not the federal government will allow for an increase in the price of natural gas, which could encourage and stimulate the search within our national territory.

World Gas Production

Table 7.2 indicates that just four nations are responsible for the production of more than 90 per cent of the world's natural gas. The United States dominates this activity as it dominated the petroleum industry in its earliest days. The United States and the U.S.S.R. have become dominant in part because the large gas fields were located within their boundaries. This made the construction of pipelines between field and market relatively easy and without any political problems.

The perfection of the technology and equipment necessary to liquify and transport natural gas across oceans will encourage the consumption of gas which otherwise would be flared off, reinjected into the oil field, or

TABLE 7.1

Natural Gas Production and Reserves by States, 1970

(billions of cubic feet)

State	Production	Reserves	Rank by Production
Arkansas	181	2,581	9
California	649	6,300	6
Colorado	106	1,629	11
Illinois	4	415	20
Kansas	900	13,325	5
Kentucky	78	978	12
Louisiana	7,788	82,957	2
Michigan	39	940	17
Mississippi	126	1,334	10
Montana	43	1,100	15 (tie)
Nebraska	6	58	19
New Mexico	1,139	13,290	4
North Dakota	35	567	18
Ohio	52	993	14
Oklahoma	1,595	16,954	3
Pennsylvania	77	1,378	13
Texas	7,853	106,353	1
Utah	43	1,065	15 (tie)
West Virginia	242	2,436	8
Wyoming	339	4,243	7
Other States	120	31,850	–
Total	21,921	290,746	

SOURCE: *Statistical Abstract of the U.S.*, (Washington, D.C.: U.S. Department of Commerce, 1972).

TABLE 7.2

The Ten Leading Natural Gas Producing Countries, 1970

(millions of cubic feet)

Country	Gross Production	Marketed Production
1. United States	23,786,453	21,920,642
2. U.S.S.R.	7,520,000	7,063,000
3. Canada	2,624,247	2,295,278
4. Venezuela	1,710,200	348,630
5. Netherlands	1,118,375	1,107,427
6. Iran	1,094,194	396,333
7. Romania	875,443	875,443
8. Saudi Arabia	710,940	104,182
9. Libya	683,900	0
10. Mexico	665,026	481,106

SOURCE: *Minerals Yearbook, Vol. I, Metals, Minerals, and Fuels*, (Washington, D.C.: U.S. Department of the Interior, 1971).

left in place. Also, the continued discovery of gas fields, such as the discovery in Groningen Province in The Netherlands, will offer natural gas in large quantities to new and waiting markets. At present, the Groningen field is suppyling the Netherlands with about 25 percent of her total energy needs, as well as providing natural gas to other European nations. Other discoveries in the North Sea area, Australia, and North Africa suggest that this is just the beginning of the natural gas industry on a worldwide scale.

reserves

As yet, little is known of the extent of proven natural gas reserves. The United States and the U.S.S.R. are considered to have 50 percent of the world's reserves, but with so many areas relatively unexplored, such a ratio could quickly change during the next twenty years. There is much to suggest that, as the value of natural gas increases, many new gas fields will be discovered.

Chapter **8**

The Transportation of Gas

Pipelines have always been the most important means of transporting gas. Problems did exist in trying to move large quantities of gas long distances. The turning point in gas pipelines is usually identified with the 1920s in the United States. This was due to three factors: the development of thin-wall, high-tensile strength, large-diameter pipe; the low cost of gas in comparison with other fuels; and the strong demand for a fuel having the qualities and applications of gas.

Gas movement was divided between the rather local distribution of manufactured gas in those parts of the United States where coal was available and relatively cheap, and the somewhat longer distances which natural gas was moved, due to its extremely low price at the wellhead. As late as 1944, manufactured gas still dominated the industry in the United States. But the end of the Second World War saw the rapid construction of more than 100,000 miles of pipeline. Today few areas of the United States lack access to natural gas supplies, and the small amount of manufactured gas is consumed principally by certain industrial users. The problem is not one of having insufficient pipeline capacity to move the gas but, rather, not having sufficient supplies of natural gas to keep the established American natural gas pipeline network working at capacity.

Pipeline Network

The map showing the American natural gas pipelines indicates their widespread coverage of the forty-eight states (figure 8.1). However, these pipelines have been built to serve the needs of specified market areas, and, if a shortage occurs in one region it is impossible, because of management decisions or the lack of interconnections, to send surplus gas to the areas in need of more fuel. It would appear to be in the national interest to

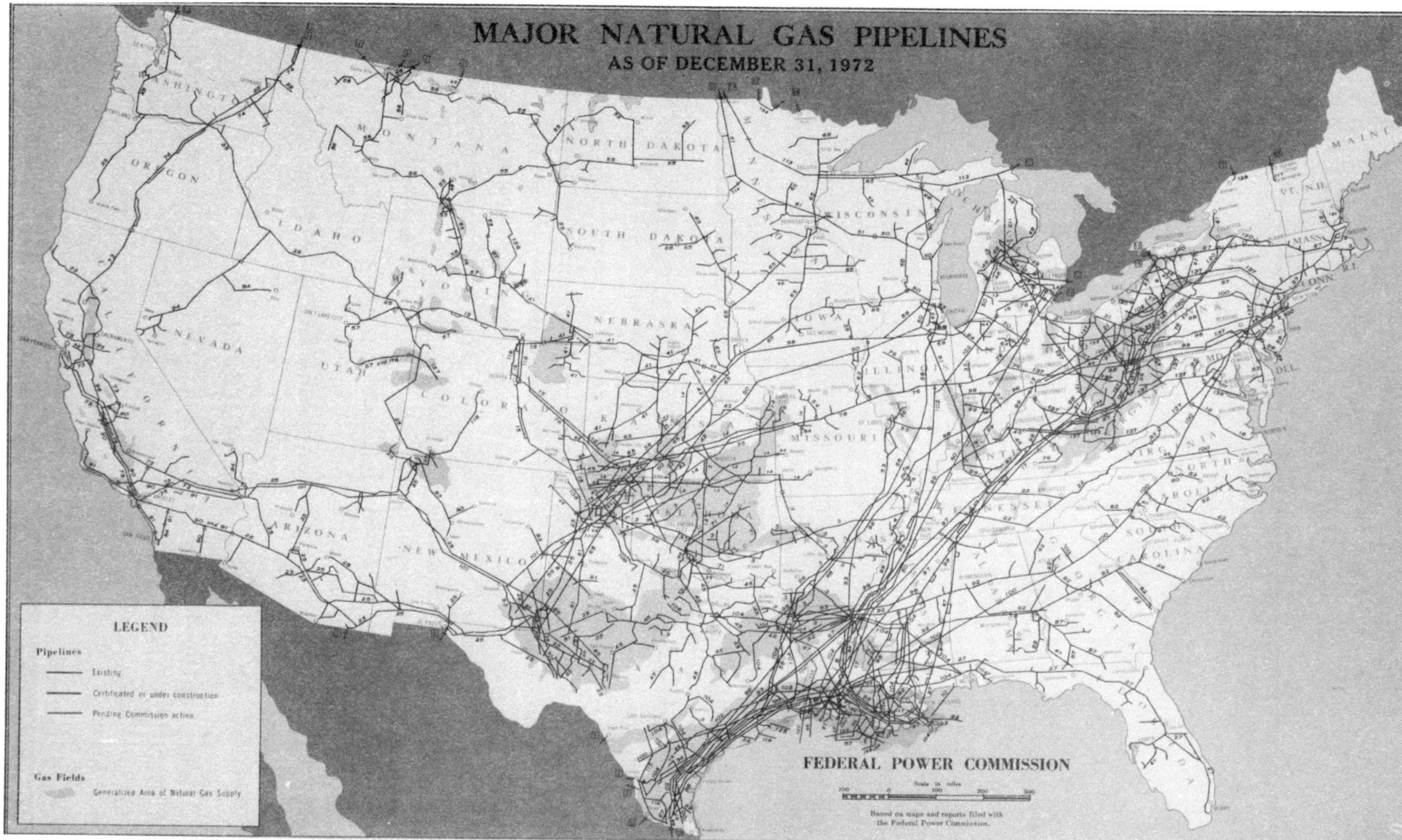

FIGURE 8.1. Gas pipelines in the United States

encourage the construction of interconnections, similar to the interconnections between electric power grids, to remedy this shortcoming.

pipeline economics

Because of its gaseous natural state, natural gas is even more suitable for pipeline transportation than oil. In the United States, pipeline construction has been relatively easy because the fields are within the national territory. In other regions, especially in the Middle East, the problem of pipeline construction is extremely difficult because of political diversity and continued strife. The U.S.S.R. is now undergoing a development similar to that in the United States three decades ago. Large natural gas deposits are being developed by pipelines hundreds and thousands of miles long, to serve markets within her boundaries.

Like oil pipelines, the major expense is the cost of the pipe and laying it underground. Therefore, it is most desirable to install pipe of as large a diameter as possible in order to obtain the lowest transportation costs. This necessitates a huge supply of gas and a sufficiently large market to absorb the transported gas. Unfortunately, such ideal situations seldom exist and the transmission companies are forced to build smaller diameter pipelines. In the United States, there were 310,000 miles of utility gas main in 1945. By 1970, this total had increased to 914,000 miles.

seasonal demands

In many of the northeastern states, the demand for natural gas is greater during a cold spell, for example, than the pipeline can deliver. It can also be stated that, during the summer months, the demand for natural gas is far below the capacity of the pipelines. To remedy this situation, natural gas producers, distributors, and pipeline companies have developed a scheme to keep all sides happy. For example, in Michigan, most of the state is served by one gas company. This company took old, played out natural gas fields, and developed them into natural, underground storage tanks. This allows the company, then, to buy extra natural gas in the summer at a discount rate, which the pipeline company offers as an incentive to get a better utilization rate from their pipeline. The lower cost for the natural gas allows the Michigan company to recover its investment and the maintenance costs of its underground storage. The existence of this storage means that, during the winter months, the flow of natural gas from the main pipelines can be augmented by additional flow from the underground storage, providing the customers with a sufficient supply of gas during peak demand periods.

This situation is duplicated in many other northern states, although some of the underground storage areas are actual caverns or underground rooms that have been created by dissolving a salt deposit with water, and extracting the solution. In principle, this is also being duplicated by the construction of huge storage tanks near major metropolitan areas, which will then supplement the supply during peak winter demand.

Ship and Barge Transport

Transporting natural gas in a ship, barge, or other transport vehicle, is expensive due to the space occupied by the gas. Natural gas can be liquified by reducing its temperature to —258° F, and by pressurizing it to a point somewhat above that of the atmosphere. The resulting liquid takes up only 1/600 as much space as natural gas, per heating unit. The first such transoceanic shipment of liquified natural gas took place in February 1959, when a converted dry-cargo ship, named the "Methane Pioneer," transported a cargo of 32,000 barrels from Lake Charles, Louisiana, U.S.A., to England. Its success led to the existence and construction of conversion plants and specially-designed ocean vessels in many parts of the world to liquify natural gas for export, and to transport the liquid to another nation, where a conversion plant turns it back into a gaseous form. Today natural gas is shipped from Alaska, Brunei, Abu Dhabi, and the U.S.S.R. to Japan; from Algeria to Spain and the United Kingdom; from Libya to Italy and Spain; and from Venezuela and North Africa to the United States. Obviously, with the shortage of natural gas in the United States, such shipments and facilities will increase, and sizable quantities of liquified natural gas will begin arriving in 1977.

To move the expected quantities of gas, fleets of specially-designed tankers, each costing as much as 100 million dollars, will have to be built. Additional plants for converting natural gas to a liquid at the receiving ports, will also have to be built. Thus, the planned expansion of capital investment to provide the natural gas for an expanding world market, will require billions of dollars in the next decade alone. Many challenges lie ahead, not only in financing such huge investments, but in locating the land-based facilities.

Inland Barges

Presently, barges are used to deliver natural gas from the Gulf of Mexico to Chicago. The expense of such barges, coupled with their rather limited capacity, have discouraged their expanded use in favor of pipelines. Their use, then, is not expected to expand very much throughout the world.

Environmental Impact

The environmental impact of natural gas transportation is minimal. Certainly the fact that the gas, if suddenly released into the atmosphere, will evaporate, presents no problem similar to an oil spill. There does exist the possibility of explosion, with damage to the immediate area. However, the natural gas pipelines have an outstanding record in regard to their safe operation. In most cases, the pipeline scar blends in reasonably well with the natural environment, to eliminate any strong objection to its presence. Likewise, the transportation of natural gas by water vessel creates little threat to the environment, other than the normal discharges and leaks that are associated with ocean vessels. Such a clean record in part explains the present preference for natural gas as a fuel by residential, commercial, and industrial users.

Chapter 9

Water Power

Early Development

The energy contained in falling water has been used for centuries. One of the earliest uses was the grist mill. Grist mills were by necessity resource-oriented, located either at a natural waterfall, or at a dam site where man could create a reservoir. The water then was usually diverted through a mill race to direct the water onto the water wheel. In the early history of the United States, the grist mill, because of its importance to the people of the region, often became the focus of yearly activities, and many towns and cities grew up from the early clustering of general store, church, post office, and the like, around the mill.

In Great Britain, the birth of the industrial revolution caused the mass development of mills along the streams and rivers flowing down the slopes of the Pennine Mountains. This was repeated in New England when the textile mills were introduced, causing clusters of mills around the natural falls and rapids, or where man constructed dams. Such towns as Lawrence, Lowell, Holyoke, and Springfield, in Massachusetts, are examples of this type of development, and still show evidence of the former dependence on waterpower.

Waterpower continued as an important energy source until the advent of the steam engine. With the abundance of coal in both Great Britain and the United States, as well as other industrialized nations, steam engines proved to be more economical and flexible. It also freed industry from being tied to the river, and allowed the location of mills at other places which might prove more logical. With the development of the electric power industry and electric motors, waterpower was doomed, and few examples remain today in the developed nations. Many of these fall in the classification of tourist or educational operations, such as the Old

Grist Mill, in Sudbury, Massachusetts, where one can buy flour ground by water-powered granite millstones.

Hydroelectricity

The development of the electric power industry may have spelled the end of direct waterpower use by mills and factories, but it opened up a new, larger market for hydropower. The first hydroelectric power plants were of necessity market-oriented. Due to the limitations of technology, it was either impossible or uneconomical to transport the electricity long distances. In many cases, the market existed nearby due to the fact that historically man has often chosen to live along river banks or at select locations, such as dam and bridge sites, or fords across the river because of the water supply, and water transportation.

As technology improved, more distant sites with great potential were developed, allowing the construction of resource-oriented power plants. Hydropower plants were very popular in the early decades of the development of the electric power industry, because, once the dam and plant had been paid off, the electricity generated was essentially free, as long as the rivers kept flowing. Many excellent dam sites, however, still remain undeveloped, awaiting the discovery of new technology which would allow for more economical transmission of the electricity over long distances.

In many underdeveloped countries, potential hydroelectric powerplant sites have lost out to conventional thermal electric, market-oriented power plants. To build most hydropower plants requires a tremendous amount of capital over a period of a few years. This must be borrowed or obtained in some way from the world money market. Because of the multitude of demands for such developmental funds, those in decision-making roles are granting smaller amounts, sufficient to construct thermal power plants in many different nations. Then, these nations have to provide the cash to purchase fuel in the following years. This, however, seems to be the popular choice for the present.

In rare situations, the potential of a site is so great, even though it is too distant from the market for economical development, that it succeeds in bringing a market to its area. Northern Norway, with its relatively abundant hydroelectricity, has been able, with government assistance, to develop an important metallurgical industry. Utilizing low-cost ocean transport, ores are brought to coastal plants to be refined with the hydroelectricity from the area, before exporting the refined metals to international markets.

In North America, a somewhat similar situation exists at the now famous Kitimat project. In 1951 the Aluminum Company of Canada began construction of a huge aluminum smelter at the head of Douglas Channel, 400 miles north of Vancouver. Kenney Dam was built, thereby reversing the flow of the Nechako River east of the mountains and creating a large reservoir. A ten-mile tunnel was dug through the Coast Mountains to carry the water to the west side where, at Kemano, about fifty miles south of Kitimat, the power plant was built. The electricity produced in Kemano is then transmitted to the smelter, where it is used to reduce alumina to aluminum.

The Problem of Scale

Scale is very important in determining the feasibility of hydropower development. The amount of electricity that will be generated each year, or the maximum installed capacity, must be weighed against the total cost. Through the 1920s in the United States, the vast majority of hydroelectric plants were small, containing one or more generators in the size range of twenty to fifty thousand kilowatts. An extreme example is the Sault Sainte Marie hydroelectric power plant, the "world's longest." It contains fifteen 480-kilowatt generators and fifty-nine 585-kilowatt generators, for a total installed capacity of only 41,715 kilowatts.

In the following decades, ever larger generators have been developed, but mainly for thermal generators. At the same time, the limited capacities of hydropower sites, coupled with technological problems and limitations, have prevented hydroelectric power plants from keeping pace with the growing size of thermal power plants. Increasingly, thermal power plants have become able to produce electricity at lower cost than many hydroelectric power plants. The net result has been the elimination of hundreds of small, hydroelectric power plants, especially after the Second World War.

Potential Hydroelectricity

One often sees tables or maps which indicate the amount of developed and undeveloped waterpower in particular states, regions, or nations (table 9.1). They usually show the amount of generating capacity that theoretically could be developed. In the United States, as well as in many other countries, much of this potential power will remain undeveloped. In the first place, the scale of many of these sites is such that the cost of the resulting electricity would be more than that produced by today's modern thermal power plants.

In the present era of increased concern about the environment, there is much interest in preserving those rivers which run largely unchecked to-

TABLE 9.1

United States Water Power—Developed and Estimated Undeveloped, by Geographic Division: 1940 to 1969

(in thousands of kilowatts)

Item and Division	1940	1955	1970
Developed Water Power[1]			
United States Total	12,075	25,742	51,952
New England	1,115	1,385	1,473
Middle Atlantic	1,633	1,789	4,264
East North Central	790	943	936
West North Central	537	962	2,728
South Atlantic	2,224	3,536	5,265
East South Central	1,270	3,576	5,224
West South Central	140	948	1,946
Mountain	1,583	3,706	6,202
Pacific	2,783	8,898	23,914
Undeveloped Water Power			
United States Total	(NA)	86,895	127,990
New England	(NA)	2,586	3,330
Middle Atlantic	(NA)	8,023	4,455
East North Central	(NA)	3,051	1,576
West North Central	(NA)	6,284	4,390
South Atlantic	(NA)	7,943	9,556
East South Central	(NA)	3,707	3,810
West South Central	(NA)	3,506	3,279
Mountain	(NA)	20,668	26,655
Pacific	(NA)	31,127	70,939

SOURCE: *Statistical Abstract of the U.S.*, (Washington, D.C.: U.S. Department of Commerce, 1972).

1. Capacity of actual installations only. Electric utilities and industrial plants, excluding pumped storage capacity.

NA Not available.

wards the oceans. Any announcement of a new dam is bound to bring much vocal opposition, which has proven successful in preventing the construction of many new dams throughout the United States. This, coupled with the increasing scarcity and high cost of money, will certainly work together to prevent the development of much of the potential water indicated on the above-mentioned maps and charts.

types of hydroelectric dams

High Dams

These comprise the most spectacular dams. They are tall earthen or concrete structures which create a high *head*, and produce large reservoirs.

They are usually located on rivers having poor *river regimens,* that is, an uneven flow of water throughout the year. The purpose of the reservoir is to store the surplus of water from the heavy run-off period in order to provide as large a volume of water as possible for a relatively constant flow throughout the year. The larger the reservoir, the larger should be this constant flow of water, and therefore a larger amount of generating capacity can be installed.

Reservoirs

The most suitable sites for reservoirs is on rivers which flow through deep canyons. This not only makes it easier and less expensive to build the dam, but cuts down the surface area of the reservoir as well, thus reducing loss of water through evaporation. Since many such reservoirs are found in arid areas, siltation is often a problem. The river flowing into the reservoir carries tremendous amounts of silt, which is then deposited on the bottom of the reservoir. Over a period of time, usually in the decades, a reservoir may become almost useless because of a reduction of its storage capacity through siltation. In the worst situations, this becomes an important factor in determining the wisdom of building hydrocapacity as opposed to building thermal power plants.

Head

High dams are very significant in producing head, which is simply the difference between the water level of the forebay—the water held back by the dam—and the water level of the after bay—the water body below the dam. The higher the head, the greater the force that can be exerted by the water, and, consequently, the larger the generator size that can be successfully used in the power plant.

The highest dam in the world is Nurek Dam in the U.S.S.R. When completed, it will have a height of 1,017 feet. The highest dam in the United States is the Oroville Dam, located in northern California, which has a height of 770 feet. The exact head of such dams is something less than these figures because the water is never allowed to rise to the top of the structure.

Low Dams

Low dams are also known as run-of-river dams. They are normally located on rivers which have good regimens, that is a strong, even, year-round flow of water. To house the generators and also to be able to direct the flow from the entire river to the turbines when necessary, low dams, ranging up to 100 feet in height are constructed.

These dams can benefit river transportation by helping to maintain minimum water depths, but they do not create reservoirs of any great significance. Because of the rather low head, such hydroelectric plants usually contain rather small amounts of installed capacity.

An excellent example of a run-of-river installation, but which is unusual because of its size, is the relatively new power plant built at the Iron Gate on the Danube River. This installation, connecting the shores of Sip, Yugoslavia, and Gura Vaii, Romania, has an installed capacity of two million kilowatts, which can produce ten billion kilowatt-hours annually.

Niagara Falls

In rare cases, nature provides both dam and reservoir. The most outstanding example of this is Niagara Falls—the dam, and the Great Lakes—the largest freshwater reservoir in the world. The result is 1,627,500 kilowatts of installed capacity on the Canadian side, and 1,954,000 kilowatts on the American side, plus 416,700 kilowatts of pumped-storage, for a total of 3,998,200 kilowatts. The demand for electricity is such that all of the water flowing down the Niagara River could be diverted to the hydropower plants, leaving the falls dry. To insure that this would not happen, an agreement was signed between both nations which establishes a minimum flow which must be maintained, essentially through the daylight hours. After 11:00 p.m., an increased amount of water can be withdrawn. Since this coincides with the time of low demand, the pumped-storage installations were added to provide peaking power for the next day.

environmental impact

The construction of hydroelectric projects has always caused great consternation to a wide variety of people. Arguments are raged as to the benefits to be gained, or the hardships and losses that will result. Today such controversy is usually summed up in what are known as "environmental impact statements." Theoretically, these should discuss all of the positive factors of the construction of the hydroelectric facility (or whatever), as well as the negative aspects. Such statements should include scientific reasoning and experience, and be devoid of personal prejudice and hysteria. Such statements must now be prepared before a license can be issued for the construction of the dam and power plant.

Negative Impact

A major problem relative to the construction of a dam and reservoir is the resultant flooding of the land. This usually involves a displacement of people living within the affected area. Often this is good agricultural land, difficult to duplicate if the persons wish to continue farming in their new locations. Such disruption of an established life pattern is seldom resolved. In addition to the human readjustment and loss of fertile farmland, man-made objects are either disrupted or forever lost beneath the waters. Roads and railroads can be cut off; schools, churches, other buildings, and even golf courses have succumed to the rising water. (Recently in the United Arab Republic, a few of the ancient Egyptian monuments were rescued from inundation by the rising waters of Lake Nasser, behind the new Aswan Dam. However, the enormous cost of moving the choicest examples to higher ground precluded saving the many other monuments and ruins of the area.)

Water Levels

In the period of years that it takes to fill the large reservoirs, the rising of the water may, at times, form large, shallow, even stagnant pools and inlets of water. In some cases on the continent of Africa, such newly created pools were sufficient to support the mass breeding of a variety of flying insects, much to the discomfort of natives, who had been displaced by the water and moved to a higher and better environment. Such shallow waters in many cases have allowed an intensive growth of weeds, whose tangled growth has choked water channels, sometimes disrupting the flow of water for irrigation and power projects.

Earthquakes

An interesting result of the concentration of so much water in such a small area has just recently come to light. Investigations have revealed mini-earthquakes, ostensibly caused by the shifting of the rock beneath, due to the weight of the water. If further research substantiates such findings, a new element must then be included in the preliminary planning for hydropower reservoirs.

Aquatic Life

For a long time the effect of dams on aquatic life, in particular the salmon, native to the Pacific Northwest, has been of great interest. The adult salmon exhibit the practice of returning to the uppermost reaches of rivers in order to spawn their eggs before dying. This problem has been solved by the construction of fish ladders, which allow the fish to swim around the dam, sometimes with less effort than was required by swimming up the natural flowing river.

Aswan Dam

One of the most interesting case studies which shows the environmental changes brought about by the construction of a hydropower project is the Aswan Dam, which was completed in 1902. For centuries, the farmers living on the floodplain of the Nile River depended on the yearly flooding of the river to provide water for their crops, and fertilizer as well, in the form of a new layer of sediment which the receding flood waters left on the landscape. In the late nineteenth century, it was decided that this practice of *basin irrigation,* which sometimes washed away crops, buildings, and even caused the deaths of many people, should be replaced by the more manageable practice of *perennial irrigation.* This would involve a dam and reservoir, which would store the water and then release it for irrigating the crops as needed. Thus the first Aswan Dam, one and one-fourth miles long and 176½ feet high, creating a reservoir 200 miles long, was completed in 1902. Although this did increase the acreage that could be watered, it led to the decline of soil fertility because most of the silt was being trapped behind the dam.

The desire to further develop both water control and power generation led to the construction of the Aswan High Dam, about twenty miles south of the first dam, which was begun in 1960. This dam, three miles long and 350 feet high, will create a huge reservoir, called Lake Nasser, which will be 300 miles long and six miles wide when it is completely filled by 1980. The dam began operation in 1968, and has extended the acreage under irrigation. Large amounts of electricity are being generated for use downstream in the region of Cairo, as well as serving to attract new industries to the city of Aswan.

The greater containment of river waters by this new dam has significantly reduced the flow of sediment down the Nile River. As a consequence, the habitat at the mouth of the Nile River delta, has changed enough to cause a reduction in the number of fish living in these waters and a resulting decline in the number of those able to earn a living from fishing. At the same time, however, the creation of the huge reservoir behind the High Aswan Dam is fostering expanded fishing grounds and opening up new employment for other Egyptians. This case study clearly suggests merely a few of the many factors involved when considering such a large installation as the High Aswan Dam.

Positive Impact

To the majority of the people, the positive aspects of hydropower projects far outweigh the negative ones. In developed nations, new reservoirs allow for increased recreational opportunities. In the United States, the reservoir associated with a hydro-project immediately causes an increase in the value of the surrounding land, which is then developed if it is in private hands. If controlled by governmental or utility authority, recreational developments on a minor scale are allowed. Nevertheless, the lake attracts a throng of boaters, fishermen, swimmers, some of whom build cabins or set up tents for camping along the new shoreline. This is considered such a normal adjunct of hydroelectric facilities today that most projects, when seeking approval of Federal Power Commission members on down to the local level, include in the plans, and stress at the hearings, the recreational land uses that have been planned. Some regulatory agencies have gone so far as to refuse the granting of the license until the utility specifically added recreation land-use plans.

As already mentioned in the case study, the creation of the reservoir often allows for the inclusion of irrigation waters. This is usually done only at large, publicly-owned dams where the multi-use concept is an important part to its justification. Hoover Dam and Grand Coulee Dam both provide sufficient water to irrigate more than one million acres of farmland, and still generate large amounts of electricity. This provides for increased agricultural production and jobs.

In some underdeveloped nations, the completion of the reservoir has provided improved transportation by allowing the people to use lake boats to travel back and forth, sometimes between areas which were formerly isolated due to a lack of land routes. In similar fashion, improved fishing and an ability to get the fish to market has affected the diet and the economy positively.

The creation of very large reservoirs may affect changes in the microclimate of the region. In any sense, it changes the physical environment in looks and introduces a new base for the ecosystem. Thus the variety of elements in the natural sphere is increased.

Pumped-Storage Power Plants

In recent years, an old technique has gained wide appreciation as a very economical way to provide peaking power for electric power systems. This technique is called pumped-storage generation. Long used in Europe, the first such plant in the United States was constructed on the Rocky River in Connecticut in 1928 (table 9.2). It had a nameplate rating of 7,000 kilowatts. But then no new similar installations were built until the 1950s when three went on-stream. In the 1960s, the American electric utilities turned to pumped-storage in a large-scale approach to help solve the problem of peak power demands (tables 9.3, 9.4). Two main factors accounted for this change. One was the increasing number of extremely large generators at new thermal power plants around the nation. The investment of such large sums of money made it mandatory to keep these conventional and nuclear thermal power plants operating at as close to capacity on a twenty-four-hour basis as was possible. The other factor was

TABLE 9.2

United States Pumped-Storage Projects On-Stream, January 1, 1970

Plant Name	State	Date Completed	Generating Capacity
Rocky River	Connecticut	1928	7,000 KW
Buchana	Texas	1951	11,000 KW
Flatiron	Colorado	1954	9,000 KW
Lewiston	New York	1962	240,000 KW
Taum Sauk	Missouri	1963	408,000 KW
Smith Mtn.	Virginia	1965	132,000 KW
Cabin Creek	Colorado	1966	300,000 KW
Muddy Run	Pennsylvania	1967	800,000 KW
Edward Hyatt	California	1968	293,000 KW
Thermalito	California	1968	82,500 KW
Salina	Oklahoma	1968	130,000 KW
Hiwassee	No. Carolina	1956	60,000 KW
Yards Creek	New Jersey	1965	339,000 KW
San Luis	California	1968	424,000 KW
Total			3,235,500 KW

SOURCE: Enclosures Number 29737, Number 51747, Federal Power Commission.

TABLE 9.3

United States Pumped-Storage Projects Under Construction, January 1, 1970

Plant Name	State	Completion Date	Generating Capacity
Jocassee	So. Carolina	1974	610,000 KW
Carters	Georgia	1974	250,000 KW
Kaysinger Bluff	Missouri	1977	160,000 KW
Kinzua	Pennsylvania	1970	396,000 KW
Northfield Mtn.	Massachusetts	1972	1,000,000 KW
Castaic	California	1977	1,200,000 KW
Ludington	Michigan	1973	1,630,000 KW
Blendheim-Gilboa	New York	1973	1,000,000 KW
Grand Coulee	Washington	1971	97,000 KW
Morman Flat	Arizona	1972	50,000 KW
Horse Mesa	Arizona	1972	100,000 KW
Elburt	—	1975	100,000 KW
Total			6,593,000 KW

SOURCE: Enclosures Number 49737, Number 51747, Federal Power Commission.

TABLE 9.4

United States Pumped-Storage Projects in the United States,
January 1, 1970

Plant Name	State	Starting Date	Generating Capacity Planned
Cornwell	New York	–	2,000,000 KW
Blue Ridge	Virginia	1976	1,600,000 KW
Laurens Shoals	Georgia	–	216,000 KW
Montezuma	Arizona	1978	500,000 KW
Blair Mtn.	Colorado	–	525,000 KW
Merrill Lake	Washington	1990	500,000 KW
Havasu	Arizona	–	1,000,000 KW
Bear Swamp	Massachusetts	1974	600,000 KW
Stony Creek	Pennsylvania	1976	1,950,000 KW
Marble Valley	Virginia	1975	1,000,000 KW
Racoon Mtn.	Tennessee	–	1,350,000 KW
Tocks Island	New Jersey	–	1,300,000 KW
Dirt Face Mtn.	Washington	–	65,000 KW
Sixth Water	–	–	90,000 KW
Upper Oconee	Georgia	–	210,000 KW
Total			12,906,000 KW

SOURCE: Enclosures Number 49737, Number 51747, Federal Power Commission.

that, for many systems, this was the cheapest method for providing the needed electricity for peak demands.

What is pumped-storage and how does it work? Pumped-storage is a form of hydropower in which bodies of water exist at two different levels (figure 9.1). The greater the difference in elevation, the better that site is likely to be, given sufficient reservoir space. During off-peak power demand, at night and on weekends, thermal power plants (usually but not exclusively) generate electricity which is sent to the pumped-storage plants. There it is used to power either conventional pumps or reversible turbines, which lift water from the lower level to the upper reservoir. Then, during the following day or week, the water can be released to flow back downhill through the turbines, to generate the needed peak power.

simple operation

The operation is relatively simple and some plants are operated remotely from a control center many miles away. These plants demand few workers. For example, the Ludington Pumped-storage Plant in Ludington, Michigan, with an installed capacity of 1900 megawatts, can operate with a work force of ten men, whereas the same capacity in a nuclear power plant would require 100 men. This plant will be the world's

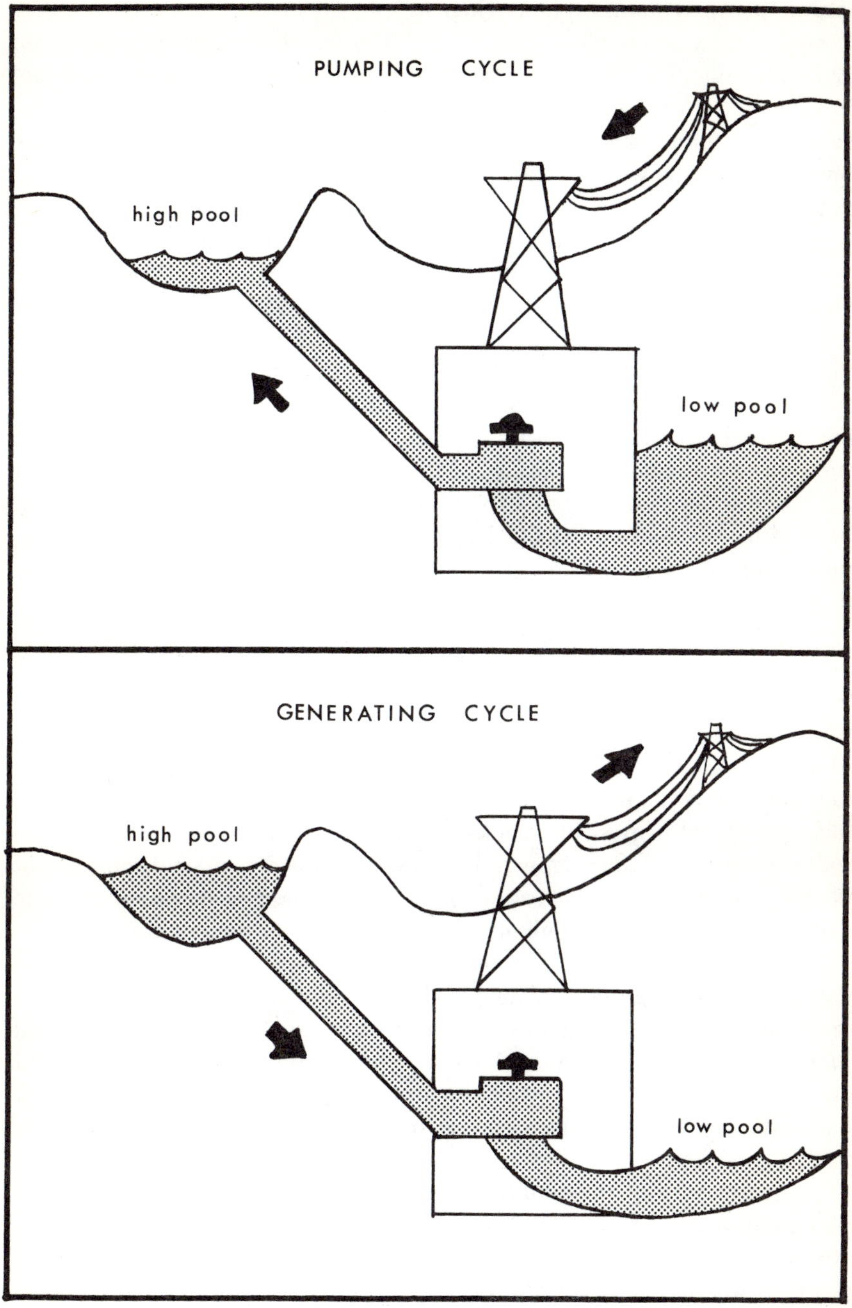

FIGURE 9.1. Operation of a pumped-storage plant

largest pumped-storage plant when it goes on-stream sometime in 1973. It contains six reversible hydraulic turbines which can pump water from Lake Michigan up through the penstocks, twenty-four to twenty-eight feet in diameter, to fill the two and one-half-mile-long, man-made reservoir, capacity of which is twenty-seven billion gallons. When all penstocks are open to allow the water to flow downhill through the turbines, the resulting flow of thirty-four million gallons a minute will be greater than the normal flow of the Ohio River as it passes Cincinnati, Ohio.

To operate the pumps at any such installation requires 50 percent more electricity than can be generated by the resulting falling water. Nevertheless, it can provide power at a lower net cost to its owners, Consumers Power Company of Jackson, Michigan (51 percent), and the Detroit Edison Company of Detroit, Michigan (49 percent), than alternative generating methods. Because the Michigan Power Pool cannot make full use of such an installation with its present size, one-third of the plant's output for the next ten years and one-sixth for the five-year period thereafter will be sold to the Commonwealth Edison Company of Chicago. This agreement enabled the owners to complete the project with resulting economies due to construction now.

On the East Coast, Consolidated Edison Company of New York has been trying to build a 2,000-megawatt pumped-storage plant at their Storm Mountain site, on the shores of the Hudson River. Although they first applied for a license on January 29, 1963, and eventually received it March 9, 1965, environmentalists and others have been successful in blocking the company from actual construction. Thus while New York City faces the threat of possible black-outs, almost yearly brown-outs (voltage reductions), and a dangerously low reserve capacity, a small number of people prevent the construction of a power plant that would provide needed electricity with minimum negative effect on the environment.

It is inevitable that many more pumped-storage plants will be built to meet the increasing demand for peaking power. Consumers Power has already purchased a 4,500-acre site about forty miles north of the Ludington Plant, near the small town of Arcadia. There are tentative plans to build another pumped-storage installation there in the 1980s. Its cost is estimated at one billion dollars for an installed capacity of approximately 6,000 megawatts.

In many parts of the country, natural sites are lacking and must be in part created by man. The reservoir for the Ludington project is man-made atop sand dunes and glacial moraines. Other more exotic plans include one proposal for the Detroit Edison Company. In this plan, it was suggested that a power plant be built beneath the City of Detroit in an old salt mine. The Detroit River would serve as the upper reservoir while large rooms of the salt mine, treated to resist dissolving and the possible collapse of the roof, would be the lower reservoir. At off-peak times, the pumps would be used to lift the water out of the mine and back into the Detroit River.

environmental impact

Pumped-storage power plants cause relatively little abuse to the environment when compared to other power plants. Yet their planned construction will often bring forth vigorous protests.

Negative Impact

One concern is fish kills. During the cycle, when such large quantities of water are being sucked in by the pumps, some fish may be killed. Another complaint is that the release of the water during the generating cycle into a river or lake will stir up large amounts of mud, thus killing certain living organisms and causing a change in the ecosystem of that portion of the lake or river. Here again, some change in the exact makeup of the ecosystem may take place, but this is a very small price to pay for a plant which will generate power so inoffensively, while in some cases providing the area with new amenities.

One cannot argue that power plants and transmission lines do not change the appearance of the landscape. Yet the low profile of hydroelectric power plants and the general neatness around such projects due to the lack of fuel shortage, tall smokestacks, cooling towers, etc., keep this problem to a minimum, and concessions have to be made somewhere with an expanding population and energy demand.

Positive Impact

The lakes created by pumped-storage projects, even though subject to periodic changes in the water level, have become recreational foci around the nation. On land adjacent to or surrounding the reservoir, picnic and camping grounds are being developed to welcome the urban dweller who is seeking a retreat to nature. Although these water bodies are man-made, the excitement, relaxation, or whatever benefits that people derive from a walk, a picnic lunch, or a weekend campout, must certainly be considered as an argument in favor of such plants. Certainly one would normally not choose these activities next to the traditional thermal power plant.

The construction of a pumped-storage installation implies that somewhere else electricity will be generated for the pumping cycle, usually at thermal power plants. Such generation suggests that base load plants are being better utilized; that such problems of thermal discharge, ash disposal, fuel handling, and air pollution are being concentrated at the larger, newer plants, which are more likely to have better control over their negative effects; and that many more power plants, with their scattered impact on resources and environment, are not built to meet peak loads.

Tidal Power

A potential source of abundant power, which has intrigued the mind of man for generations, is tidal power. The range between the level of water at high tide and at low tide may be as much as fifty feet. Tidal power can be harnessed using essentially conventional hydroelectric power-plant technology. A major problem, however, is that throughout the month, the tides vary between their maximum level (spring tides) and their minimum level (neap tides). Furthermore, the time of the high tides is continually changing, so one can never depend on generation at a given hour of every day.

The number of high and low tides varies from coastline to coastline. Most coasts experience two high tides and two low tides during a twenty-four-hour period. Some areas have only one high tide and one low tide, while other areas, such as the Mediteranean Sea and the Gulf of Mexico, experience modest tidal ranges.

sites

The best sites are located on estuaries or narrow inlets, where the tides reach the maximum range of forty to fifty feet. This facilitates the task of building dams containing water gates and power plants. Because of the topography of certain sites, more than one dam can be used to impound a larger body of water. This is the case with the existing tidal power plant in Brittany, France, and the proposed plant on the Bay of Fundy.

The United States and Canada have considered different proposals for the construction of a joint project utilizing the tidal range characteristic of the Bay of Fundy in Maine and New Brunswick, but these have been rejected for a variety of reasons. Across the Atlantic, the French began construction of the world's first tidal power plant in 1961. It is located at the mouth of the Rance River, in the Gulf of St. Malo, Brittany. The project went into operation in 1966, utilizing the tidal range of up to forty-four feet, to generate electricity which is then fed into the French grid. Today controversy still rages as to the economic viability of the project.

environmental impact

Very little can be said against tidal projects. The dams they build do effect the movement of some members of the ecosystem, which might formerly have moved unhindered in and out of the dammed estuary. The concentration of the outflow of water may change or intensify coastal currents but more than likely will cause little damage, especially when one considers what the ocean itself can do to the coastline during a storm.

As with any power plant, some would object to its construction on the seacoast. Again, careful design and location may make it possible to keep it from view, except in the immediate area. Such plants produce no air or thermal pollution and certainly on this basis should be welcomed as a means of supplying needed energy. Unfortunately, experience gained thus far indicates that technological difficulties still exist. As a consequence, few tidal power plants will be built during the coming decades.

Chapter 10

The Transmission of Electricity

Improvements in technology are extending the distances over which electricity can be transported economically. Originally, electricity was generated within a mile or so of the point of consumption. But, with the design of improved cables, insulators, towers, and the knowledge of the advantages of higher voltage levels, electricity now is being transported more than 1000 miles. For the majority of transmission, however, these distances average less than 500 miles.

Water Power

At first, man used the energy contained in falling water in a mechanical way. This energy was transferred from water to machines via a drive shaft. The effective length of this shaft limited the utilization of this energy to some tens of feet. Thus, early water-powered factories and mills were long and narrow, built parallel to the river, and often multi-storied, because of the ease and relative efficiency with which this power could be transferred to upper floors. This all ended with the discovery of electrical generation, for hydroelectricity was more flexible and could be transported far from the edge of the river.

The earliest long-distance transmission lines were associated with hydroelectric power plants. In 1913, the Big Creek transmission line, in southern California, extended 240 miles into the City of Los Angeles, and operated at 150 kilovolts. This was later raised to 220 kilovolts in 1923. In the 1930s, Los Angeles built a 287-kilovolt line to bring in electricity from Hoover Dam, about 300 miles away. No additional, extra-high, voltage lines were built until the early 1950s, but, by then, more and more of the power transmitted was thermally generated. Today, the vast majority of transmitted power comes from conventional, or nuclear, thermal power plants.

types of transmission lines

Transmission lines can be defined as those lines which are used for the bulk transmission of electricity between a generating (receiving point) and major substations (delivery points). They can be either high voltage or extra-high voltage lines. Extra-high, voltage transmission lines are those rated higher than 230 kilovolts. Practically all transmission lines in the United States are alternating current (A-C) lines.

Transmission Towers

To carry these lines across the countryside, towers must be erected. These can be classified as "t," "h," or self-supporting and guyed towers (see figure 10.1). The "t" is the simplest design, with one pole and cross-bar. When made of wood, it can support the lightest cables designed for the lower voltages. The "h" design is also usually made out of wood, using two poles and one cross-bar. Because of the added strength, it can support heavier cables and therefore higher voltages. It also requires a wider right-of-way. The self-supporting and guyed towers are the tallest and strongest, capable of carrying the heaviest cables and the highest voltages now employed. Obviously, towers can be built to different sizes, capable of varying amounts of stress, and they are designed according to the terrain and climate, as well as to the weight of the cable and the voltage level.

Right-of-way

To construct a transmission line, a right-of-way must be acquired. This is a time-consuming and expensive proposition in many countries and today is facing the added challenge of organized opposition from home-owners' associations, city governments, and environmental groups. In the United States, the cost is rising rapidly, routes are being altered due to courtroom decisions, and utilities are placing greater effort in planning for, and constructing, new transmission lines.

The larger the tower, the wider the right-of-way, and, consequently, the higher the land costs. Thus a utility attempts to project the maximum amount of electricity which will flow through a given corridor. On that basis, one or more circuits are planned, using a single row, or parallel rows, of towers.

Alternating Current and Direct Current

Electricity can be transported as either alternating or direct current, better known as A-C or D-C. In the United States, practically all of the

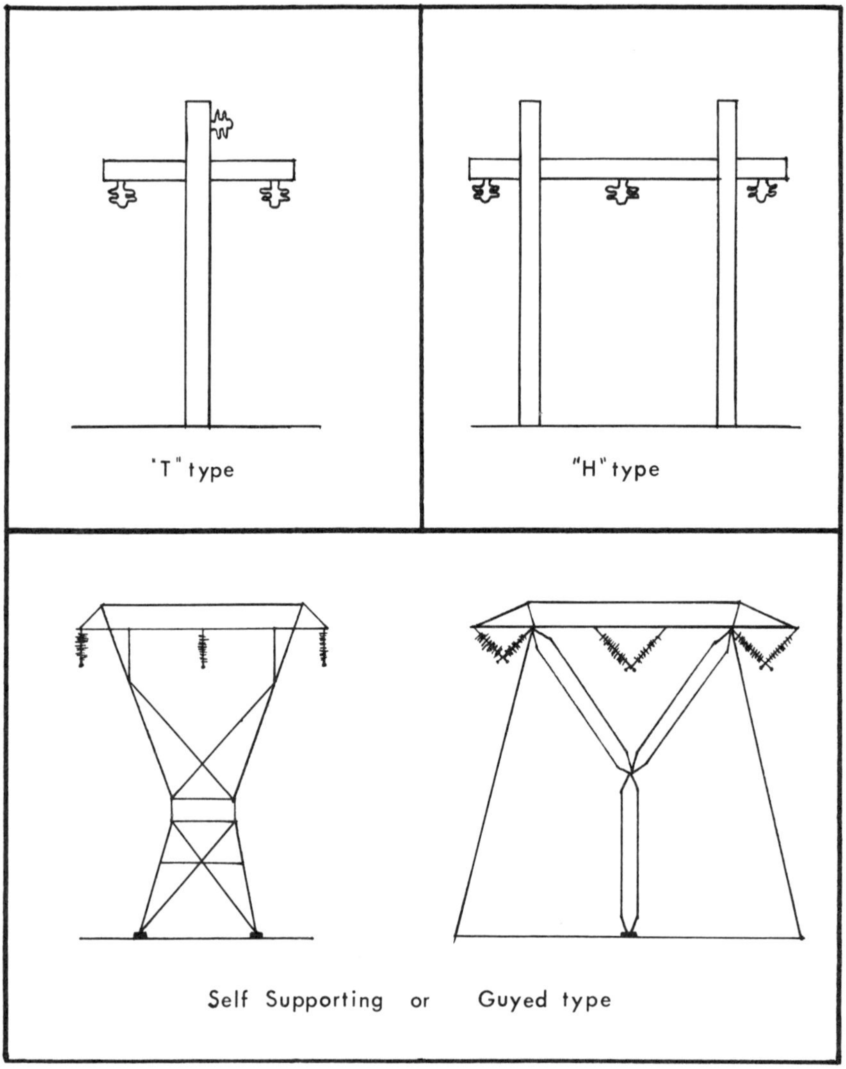

FIGURE 10.1. The three main types of transmission line towers

transmission lines are alternating current, but this was not always so. Early in the history of the electric power industry, Thomas Edison opted for D-C lines, and used them in his system. Competing electric utilities used A-C lines, and these became the accepted norm for the American industry.

A typical three-phase A-C circuit is composed of four or five cables. Three of these cables transport the electricity, while the fourth and fifth, often smaller-diameter wire, function as a lightning shield. The one or two wires normally are higher than the A-C cable, and usually intercept the lightning before it hits the circuit with destructive results. Since these protecting wires are grounded, the force of the lightning is sent harmlessly into the ground at the base of the pole or tower.

With direct current, the electricity does not change directions back and forth along the circuit, as with alternating current, but rather flows in one continual direction. A D-C circuit can be operated using only one wire and allowing the ground to act as a return. In the United States, and many other countries as well, two wires are used per circuit, one wire carrying the current away from the power plant and the other carrying it back towards the power plant. High-voltage D-C line costs, excluding the terminal equipment, are substantially less than for an equivalent overhead A-C line, because of reduced right-of-way requirements, fewer conductors, and lighter, cheaper towers. Such D-C lines can be built for about 65 percent of an equivalent A-C line, but the great expense in building the terminal structures means that D-C lines can be competitive only if used over long distances of hundreds of miles, or under unusual situations, such as a submarine cable. In 1954, the first commercial underwater D-C line was built, connecting the Island of Gotland with the Swedish Mainland. This was followed in 1961 by a cable beneath the English Channel, which connects England and France. Several others have been built more recently.

In the United States, the only high voltage D-C transmission line is the 851-mile-long NW-SW Pacific Intertie. It connects the John Day hydroelectric power plant on the Columbia River with Los Angeles, California. The line has an appropriate rating of 1440 megawatts. The Sylmar converter station, built near Los Angeles, required twenty-one acres of land, or almost 50 percent more than for a similar A-C station. Thus increased land costs are just one factor in the high cost of D-C terminal facilities.

Land Acquisition

Formerly, the task of building a new transmission line was relatively simple. From on-sight inspections and aerial photographs, a route would be chosen for the new circuit. The route would be chosen so as to avoid swampy land, subdivisions, etc., while looking for land which would be easy to build towers on and to maintain the right-of-way. The shortest route also was important because of its lower construction cost. Landowners, who resisted the intrusion of a transmission line, often lost their appeals in the courtroom and received lower payments than had been offered by the utility before taking their case to court.

Today, it's a different ballgame. The average citizen is winning some decisions, not on simple opposition to new transmission lines, but in favor of less offensive locations. New routes have been adjusted away from areas built up with homes, out of sight of national monuments and parks, and around commercial areas. Utilities are now paying special attention not only to the location of the new route, but to the materials used in the towers. Instead of the typical slant across a highway, with the greater exposure of barren right-of-way, new lines are often crossing highways at right angles, which reduces their offensiveness. Towers are chosen with regard to both material and color, so as to blend more easily into the surroundings. There is a definite emphasis on locating new lines in valleys, or on the sides of hills, so there is less interruption of the horizon by the towers and cables. Certainly there is a great need for coordination between each utility's management and the people living in its service area, in order to insure delivery of the electric energy desired while retaining as beautiful an environment as possible.

Another problem area involves the cities. As the cities have increased in size and the suburbs have expanded with urban-related growth, it has become increasingly difficult to transport the vast amounts of energy required in the city by commercial and industrial users particularly, along the existing right-of-ways. One solution is to tear down existing towers and replace them with larger structures, capable of carrying two or three circuits. Unfortunately, in many such situations, it is impossible to bring the electricity normally carried by that circuit into the city from another route, and the utility simply cannot shut it down. Thus, the utility must build a parallel line if they can obtain the needed right-of-way (in the event that the present strip of land cannot accept it), or they must seek a new route. In many cities, the decline of public transportation has proven an asset to utilities with such a problem. Railroads are selling or leasing land along their right-of-ways to the utilities, for the construction of high voltage transmission lines.

The answer, of course, which everyone is hoping for, will be the perfection of efficient, economically priced, extra-high voltage transmission lines which can be put underground. In the meantime, the struggle goes on as all sides try to choose the best location for power corridors.

Voltage Levels

In designing a transmission line, it is highly desirable to construct as high a voltage circuit as is feasible. It is not only cheaper to build, but it moves the electricity at a lower cost. Unfortunately, it is neither wise nor possible to construct the towers for their maximum capacity. The extra cost, when averaged over the many years during which only one circuit hangs on the towers, or the inefficiency of operating a circuit at a fraction of the rated voltage, would eliminate the expected economies of operation.

Coal-by-wire

Moving electricity by transmission lines costs money. Transporting coal or lignite by train also costs money. In some instances, economic studies indicate that it would be more efficient, that is, cheaper, to construct mine-mouth power plants and transport the electricity generated by an extra-high, voltage-transmission line to the market. In the southwestern part of the United States, several power plants have been built which exemplify this situation. The electricity thus generated from existing plants (the Four Corners plant, as such, is the most infamous plant) is sent hundreds of miles by an extra-high, voltage-transmission line to such markets as Phoeniz, Arizona, and Los Angeles, California. Present plans call for an increase in the number of such plants not only in the Southwest, but in other parts of the United States.

Little objection is heard from those consuming this energy in the distant cities, but those living closer to the plants, as well as those who love the natural beauty of the Southwest, are attacking such development on the basis of air pollution, thermal pollution, and desecration of the landscape by the necessary transmission lines, to mention a few negative factors. Those plants now planned will be completed, but additional power plants will face more vigorous opposition.

Interties

The increasing size of generating units being built by the electric utilities has encouraged the construction of interties—interconnections between such utilities. This has been done as one way to insure a supply of electricity during emergencies, as well as during times of regular maintenance on plant equipment. Some utilities prefer to purchase their peaking power via interties rather than construct their own plants for such purposes. This usually results in economies to all utilities involved.

Joint Purchases

The use of interties has also led to the cooperative purchase of new generating units. This allows both utilities to purchase one large unit such as a 500-megawatt generator instead of each buying a 250-megawatt unit. The savings that result from the construction of just one unit, plus the production of lower-cost electricity by the larger unit, easily offsets the cost of building an intertie if one does not yet exist. In these times, when it is becoming increasingly difficult to find a plant site which will receive approval for power plant construction, one system may have many sites, or room for expansion, at existing plants. By the use of interties and joint ownership of plants, the most economical, least harmful-to-the-environment sites may be developed. In a somewhat similar way, many small private and public utilities, faced with increasing problems of financing new construction, have built interties to larger, adjacent electric systems and are relying on another system for peaking power, as well as power needed for the increased demand for electricity in their service area.

Exchange of Electricity

Some utilities have a surplus of power during the winter or the summer, and interties allow for the sale or exchange of this power with other utilities in need. Some utilities have large amounts of surplus hydropower at night and on weekends, and via interties this power can be used by nearby utilities, which can then reduce their fuel costs by shutting down some of their conventional thermal units. Utilities that are located in different time zones have discovered the advantage of exchanging power to meet peak demands that occur one hour apart. The sale of surplus power from the Ludington Pumped-storage Plant to Commonwealth Edison Company in Chicago is in part due to the time difference between peak demand in Michigan and Illinois.

National Grid

In the United States, the ultimate of this technique would be the construction of an extra-high, voltage-transmission grid, either A-C or D-C, that would connect all the major utilities from coast-to-coast. Thus utilities in the Pacific, Mountain, and Central Time Zones would send their surplus power towards the East to help meet the six o'clock peak demand in the Eastern Time Zone. As the peak demand moved westward, the surplus power in the unaffected time zones would be shifted appropriately. Not only could savings be realized from such cooperation to supply peaking power, but such a grid would greatly enhance the reliability of systems and further reduce the occurence of another Northeast Blackout.

Perhaps the biggest hurdle facing the construction of such a grid is that of ownership. Who should build it and who should control its use? Its greatest weakness is the lack of similar-sized markets with the appropriate amount of installed capacity in each of the four time zones.

When thinking of national power grids, the U.S.S.R. must be mentioned, for this nation is crossed by eleven time zones. But their electricity production and consumption is even more unevenly distributed than in the United States. While the discussion goes on concerning national power grids in the United States and the U.S.S.R., American utilities are forming their own power pools and regional grids from which the public benefits.

Environmental Impact

Positive Impact

A very strong case can be made in favor of transmission lines if one thinks only of the physical environment. The right-of-ways, when kept clear of underbrush and trees, have, and can, function as firebreaks. The roads formed during the construction period, and used for maintenance, can provide access into areas for recreation or economic development—logging and ranching, for example—and the land beneath the transmission lines can be farmed or used for grazing.

Negative Impact

A cleared right-of-way can make an ugly slash across the countryside. The use of chemicals to retard or kill vegetation produces a corridor which is in sharp contrast with the surrounding landscape. Access roads may also look unattractive and contribute to erosion, especially on hilly land. But perhaps the major disadvantage is that tall towers carrying cables are simply ugly. They disrupt the eye when looking across the land, and, although painting the towers appropriate colors helps, most people still mentally object to their existence and construction.

Chapter 11

Uranium and Nuclear Power

Uranium is the main source of atomic energy. It is found throughout the world but only in a few ores is its concentration great enough to allow production. The richest uranium ore is pitchblend. The two most famous deposits are in Zaire and near Great Bear Lake in northern Canada. These areas were the only major producers before World War II. Because of the demand after the war, many new deposits were developed, especially in the United States (table 11.1). Carnotite is the most common ore in this country.

World Production

The United States produces almost one-third of the world's uranium. When that amount is added to the production of Canada and South Africa, it accounts for almost two-thirds of the world's uranium production, if estimates are correct for the Communist nations. As the demand for uranium increases, other ores will be utilized and the list of producing nations can be expected to expand.

united states production

In the United States, the major deposits are found in a belt extending from Arizona and New Mexico to Montana. New Mexico is the leading state, producing more than twice as much as the second state. The ores are not especially rich but large-scale production techniques keep costs relatively low.

other nations

In Canada, the deposits lie in a broad belt extending from North Central Canada near the Arctic Circle to Ontario. Ontario is the leading

TABLE 11.1

Major World Reserves of Uranium, 1965

(based on economically recoverable reserves at mid-sixties prices)

Country	Tons of U_3O_8 (An oxide that contains 84.8% uranium)
1. Canada	200,000
2. United States	150,000
3. Republic of South Africa	140,000
4. France	35,000
5. Other Nations not Communist	50,000
Total	575,000

SOURCE: *Statistical Abstract of the U.S.*, (Washington, D.C.: U.S. Department of Commerce, 1972).

producer in Canada with about 75 percent of Canada's total production. In Zaire, the Shinkolobwe mines used to produce almost pure pitchblend, but they have just about exhausted these ores and are now mining lower grade ores. In South Africa, the uranium is a by-product of the gold mining industry. As with many other mining operations, the processing of the ore to get minerals other than the primary metal often helps make the whole mining venture profitable and, in this case, results in another product which can be exported.

Converting Uranium to a Fuel

Unlike coal or oil, uranium does not possess an inherent ability to be directly used by man because of its low concentration in the ore. Thus it must be processed to concentrate the desirable elements into what is called a nuclear fuel. This fuel is composed of a mixture of U-235 and U-238 packed as uranium oxide pellets into rods made from stainless steel or zirconium alloy. These rods are about one-half inch in diameter and up to fourteen feet long. Many of the rods are put together to form a subassembly, and several subassemblies are spaced close enough together in the reactor to sustain a chain reaction. One reactor many contain hundreds of thousands of pounds of uranium oxide.

The nuclear power plants in operation today obtain their energy by the fission process. This takes place within the reactor vessel at a carefully regulated rate. Control rods, made from cadmium, are slowly withdrawn from the subassemblies allowing the fission process to take place at an increasing pace. The fission process releases energy which is used to produce steam to drive a turbine which in turn powers an electric generator. To reduce the amount of steam being produced, the cadmium control rods are dropped back into the subassemblies, thereby reducing the rate of fission. Thus the essential components of a reactor are (1) the fuel which fissions to release energy for the production of steam, (2) the control rods and associated equipment, used to govern the rate of fission, and (3) the cooling fluid which removes the heat being generated in the reactor.

It is of interest to point out that a nuclear power plant differs little from a conventional coal-burning power plant. The nuclear reaction is providing the heat to produce steam to drive the turbines, instead of obtaining it from burning coal, oil, or gas. Thus, the efficiency of the plants in converting energy to electricity is at best about 39 percent. To make a choice between conventional and nuclear power plants, it is necessary to consider construction costs, fuel costs, availability of cooling water, environmental restrictions, and especially the size of the plant. Nuclear power plants gain their cost competitiveness by increasing size that ranges in the hundreds of thousands of kilowatts.

Environmental Impact

A major concern with the use of nuclear power plants is the increased demand for cooling water. Using nuclear fuel demands lower steam pressure and the less efficient use of steam produces more waste heat. As a general rule, nuclear power plants produce approximately 10,000 BTU per kilowatt-hour versus the 6,000 BTU of conventional thermal plants. Many of the new nuclear power plants are being required to use cooling towers or cooling ponds to reduce the temperature of this water before discharge back into a lake, river, or ocean, or for reuse at the plant. This can greatly reduce the problem associated with thermal pollution by nuclear power plants.

To some, radioactive pollutants are more of a threat than thermal pollution. Three kinds of radioactive pollutants are produced by a nuclear power plant: solid, liquid, and gaseous. The solid wastes include clothing worn by workers, tools, and other equipment used to operate and maintain the power plant. Some radioactive liquid wastes are produced due to impurities in the cooling water and their reaction to escaping neutrons from the core area. Gaseous and liquid wastes are produced due to minute cracks or leaks in the fuel subassemblies. Although the above can create serious problems, of major concern today is how to dispose of the radioactive wastes resulting from the spent fuel. The fission products build up to the point where the chain reaction stops. At this point, the power plant must be shut down in order to replace the worn out subassemblies with new ones. The old subassemblies are then shipped back to a reprocessing center in cooled and shielded containers. Here the fission products and the unreacted fuel are separated. The unreacted fuel can be used in another subassembly but the fission products must now be disposed of safely. Although some of the material is dumped into deep trenches in the oceans, after being packed in special vessels, it appears that the majority of this material will be stored in underground vaults dug in salt layers, deep beneath the earth's surface, in Kansas, New York, and the Gulf Coast area.

Radioactivity in wastes from the operation of nuclear power plants will continue to haunt the efforts of those who champion the cause of the atom. Three general practices have developed in handling the situation, (1) dilute and disperse—used for large volume, low activity wastes; (2) delay and decay—satisfactory for medium activity wastes; and (3) concentrate and contain—applied to high activity wastes which are stored underground. Radiation problems also exist in the tailings piles at uranium mines. Although this might be considered minor, it has developed around some mines that such waste material has been used to pour cement slabs, build cement block walls, and even to be used in sand boxes. In some cases, the material has been washed into streams and rivers. High radioactive counts have been recorded, but as yet no serious consequences have developed. The situation does lead one to speculate on endless implications. Radiation continues to be a problem of nuclear power plants due mainly to the challenge of disposing the fission wastes, and the concern over safeguards which would prevent massive contamination of the area around the plant.

To provide fuel for nuclear power plants requires far less mining than to duplicate the energy needs by means of coal-burning power plants. That certainly appeals to many who decry the strip mining and subsurface mining for coal in this country. The fission process does not produce the fly ash, sulfur, and other gaseous and particulate matter thrown up into the sky by way of smokestacks all over the country which have changed the environment in a negative way. If calculations are correct, electricity generated by nuclear power should cost less than that obtained by burning fossil fuels, especially in light of the increasing costs of foreign fuels. Thus, management is faced with the delicate task of choosing between energy sources which in one way or another threaten our environment as we know it today.

Chapter 12

Geothermal Power

Man has used geothermal energy, in one form or another, for centuries. Perhaps the earliest use was at hot water spas in Europe and Asia. People frequent such spas, which are still in use today, to soak in the hot water or to take hot mud baths. In Iceland, hot water has been piped into homes and other buildings to furnish heat during the long winters, since 1925. In the United States, millions of tourists over the past century have visited Yellowstone National Park to see Old Faithful erupt on schedule. More recently in the United States, much interest has been focused on the potential power that could be obtained from harnessing geothermal energy.

Types of Geothermal Power Resources

There are three types of geothermal resources—steam, hot water, and hot rocks. The steam resource can be subdivided into dry and wet steam deposits. The majority of such deposits are wet. Dry steam is preferred because it can be used directly, is therefore less costly to develop, and leaves no deposits of minerals or salts. Wet steam must be separated into steam and water before generation can take place. In New Zealand, this water is simply dumped into the streams, but in areas such as the Salton Sea in California, the high percentage of dissolved solids prevents such disposal. As a result, most plans for geothermal power plants will include provisions for responsible disposal of the mineral-laden water by reinjection, or treatment, for the removal of the minerals, thus producing water for industrial, municipal, and agricultural use.

Hot water is simply ground water which has come in contact with hot rocks beneath the surface of the earth. This water contains varying amounts of dissolved solids, mainly salts, which must be disposed of in some manner. The alternatives seem to be to sell them for commercial use when possible, or to pump the solids, dissolved in a new solution, back into the ground for recirculation.

Hot rocks can be found, at varying depths, anywhere beneath the earth's surface. Such geothermal heat is produced by the decay of radioactive materials. This heat radiates and is conducted toward the surface of the earth, thus explaining why temperatures increase as one drills a well, or descends into a mine. What appears to have the greatest potential, however, are the really hot spots, often located in a region containing a relatively recent intrusion of molten rock into the mantle, but not extruding out onto the surface. Such formations are sometimes indicated by the presence of hot springs at the surface. It has been estimated that temperatures of 572° F. can be found within 6,000 meters of the surface in much of the western United States.

Development

Electricity is currently produced commercially using geothermal energy in a few countries, while several other nations are engaging in active research. The oldest such installation is located at Larderello, in the Province of Tuscany, southwest of Florence, Italy. This dry-steam field first generated electricity in 1904. Its use is still expanding and today it has almost 400,000 kilowatts of installed capacity. In 1950, the Wairakei field in New Zealand was developed to produce electricity and supplements the country's major source of electricity, water power. In the United States, only one site has been developed, although prospecting continues for new sites in the Southwest that will justify development. In 1922, a geothermal deposit at The Geysers, located eighty-five miles north of San Francisco, California, was tapped. There was sufficient pressure and temperature in this dry steam deposit to operate a turbine, but it corroded the pipe then in use. With the abundance of cheap hydroelectricity on the West Coast, it could not compete and the project was abandoned. In 1956, interest in the project was revived with the result that four investor-owned companies joined forces to tap this resource. Union Oil Company operates the project and drills the wells for its partners, Thermal Power Company of San Francisco, and Magma Power Company of Los Angeles, California. The steam produced by the joint efforts of these three partners is then sold to Pacific Gas and Electric Company. It is their responsibility to construct and operate the generating plants and sell the electricity. In 1960, the first power plant rated at 12,500 kilowatts went on-stream. By 1968, the installed capacity was 82,500 kilowatts, with more than 600,000 kilowatts projected for 1975.

Drilling activity has increased in the United States, mainly in the southwestern states. This is centered in the southern part of the Imperial Valley. Wells have discovered wet steam and hot water, but with a high percentage of dissolved salts, perhaps 3 to 4 percent. This may make development more difficult and expensive, but the prospects look bright. One reason for this optimism is that not many miles away, on the Mexican side of the boundary, a 75,000-kilowatt, geothermal power plant is ready to go on-stream, utilizing natural steam from wells drilled in the Cerro Pierto field. This will then become the second such plant in Mexico, after the Hidalgo plant which has operated for several years.

A major deterrent to geothermal development in the United States has been the fact that many such sites in the West are located on federal lands. This problem has been somewhat relieved by the passage of the Geothermal Steam Act of 1970 by Congress. This bill directed the Geological Survey to map all potential geothermal sites on federal lands, and offer them for lease. The mapping is almost completed, and leasing will soon be possible.

Reserves

Estimating the amount of geothermal reserves in the United States is quite a challenge. The experts in this field disagree by a wide margin. Dr. Donald White of the U.S. Geological Survey suggests a conservative estimate of 30 million kilowatts just in the West, almost 10 percent of the total installed capacity of 350 million kilowatts. Others now indicate that The Geysers and the Imperial Valley may eventually yield 40 million kilowatts. Certainly other nations around the world may turn to their geothermal resources and develop them, using the experience gained in Italy, New Zealand, the United States, and most recently, Mexico.

environmental impact

Negative Impact

A major problem of natural steam wells is the noise. When the steam is diverted from the turbines and released into the air, the roar can be deafening, even with huge mufflers in place. Further technological development will most certainly provide a solution to this noise problem. Present plants are located in remote areas and, thus, present no immediate problem.

The dissolved solids in the steam and water, when separated, can create a disposal problem. The release of these solids could contaminate water supplies and watersheds if not controlled. It is estimated that a one-million-kilowatt plant, using hot water containing about 2 percent dissolved salts, would produce 12,000 tons of salt each day.

As with other developments which remove resources from beneath the surface, subsidence can be a problem with geothermal sites. However, the problem of disposal of dissolved solids, by reinjecting the solution back into the ground, will probably prevent serious subsidence problems.

Geothermal steam can contain impurities other than salts but these are found in minute quantities. The list includes methane, nitrogen, hydrogen, argon, ammonia, carbon dioxide, and hydrogen sulfide. The only one that creates pollution is the hydrogen sulfide. This dissolves easily in water and evaporates into the air during the cooling process. This is what causes the sulfur odor at The Geysers. Estimates now suggest that the amount of released sulfur might range from a release equivalent to that from The Geysers—from a low-sulfur, oil-fired thermal power plant, to slightly more than that released by an equivalent-sized thermal plant burning high-sulfur oil. Again, due to the fact that the American and Mexican plants are in remote areas, this is not now a serious problem. However, with the proliferation of geothermal power plants, anti-pollution devices will most certainly be required.

Thermal pollution exists but, thus far, does not seem to pose much of a threat. Cooling towers are used at The Geysers, and most projects will release the waste heat into the atmosphere. Much depends on the geography of the region, which could be most forgiving in absorbing the heat and moisture with little effect on the environment.

And finally, blowouts or the threat of blowouts must be considered. Since some geothermal sites will be located close to faults, there is the possibility of some problem. This however, seems to be a minor criticism.

Positive Impact

Geothermal power plants pose, perhaps, the least threat to the environment. Little disturbance of the ground surface, no visual air pollution, and no threat of radiation are just three of its advantages. Its ability to produce electricity at low cost from steam or hot water provided by nature will certainly get increased support from environmentalists who are distressed over power alternatives. Increased government and private research will hopefully allow for the probing of identified thermal anomolies, those areas having higher temperatures than the surrounding region; and this could help ease the squeeze on current energy supplies.

Chapter 13

Solar Power

Solar power is the most abundant form of energy available to man. It has been estimated that the energy arriving on 0.5 percent of the land area of the United States is greater than the projected energy needs of the country in the year 2000. Such potential has always been of interest to man, but most research has pursued other energy sources with perhaps easier challenges. However, the deterioration of the environment and the lack of sufficient energy supplies within the United States have been instrumental in causing the focusing of much attention, again, on solar energy.

Development

Though few practical devices have been constructed during the past century throughout the world, each gave hope of the great potential of solar energy. A primary challenge is the intermittent nature of this energy. This warrants a system which will somehow store the energy for use during the nighttime or during long periods of cloud cover. A second handicap is the fact that the sun's rays are diffuse. In the low latitudes, the rays are close to being vertical, but towards the higher latitudes, the same ray is forced to light and heat a larger portion of the earth's surface. Thus, in order to utilize this solar energy, the rays of the sun must be concentrated in some way, in order to intensify the heat. The hang-up, then, is in designing apparatus that can collect and concentrate the sun's rays, in order to produce heat that is economically competitive with conventional thermal sources, and yet provide ways to store this energy for use during nonsunshine hours.

Solar Stills

One successful development in the use of the sun's energy has been the solar still. These structures include a glass or plastic roof, through which the sun's rays pass. The rays are then trapped by a variety of methods, using the greenhouse principle. The heat is then used to evaporate impure, salty, or simply soil moisture. This moisture is then deposited on the inside of the slanted roof as droplets of water, which then run downslope to a collection trough to be directed to a container. This produces small amounts of pure water at a low cost, using the free energy of the sun, approximately one gallon of water per day per ten square feet of collection area. This average yearly figure is much higher during the summer months when demand is greater. The largest number of such stills have been built on the Greek Isles, long known for their lack of fresh water. Similar installations exist in Spain and Australia, and one such still, built in 1872 with a collecting area of 51,000 square feet, produced 6,000 gallons of water per day for forty years. Today, a one-acre still, with roof-like glass for collecting the condensate, should produce 5,000 gallons per day. Presently, a 9,000-square-foot pilot plant, modeled after the first experimental model of a multiple-effect still constructed at the Solar Energy Laboratory at the University of Arizona in Tucson, is being operated by the Mexican government at Rocky Point, Sonora, Mexico.

Solar Furnaces

Most people, at one time or another, have experimented with magnifying glasses by concentrating the sun's rays to set a leaf on fire or to burn their name into a piece of wood. This same principle is used in the solar furnace, except that the scale is much larger. By use of lenses or mirrors, the heat of the sun is concentrated and directed into the collectors, where there exist extremely high levels of heat, equal perhaps to 1,000 kilowatts. France has two large solar furnaces at Mont Louis in the French Pyrenees; Japan has one large furnace erected by Tohoku University in Sendai; and the United States has built one at its Quartermaster Corps Laboratory in Natick, Massachusetts. Their major use is to study the properties possessed by materials at very high temperatures.

Solar Water Heaters

Man has been building water heaters, which use the energy from the sun, for nearly a century. It has been only recently that they have been mass-produced and marketed successfully, in competition with fuel-burning and electric water heaters. Although a variety of designs exist, such heaters collect the heat from the sun during the daytime, circulating the waters so that, by the end of the day, a heater with a collecting surface of thirty square feet could heat forty gallons of water. This technique is especially applicable in those cultures where all the hot water is used for baths and dishwashing at the end of the day. In the evening, the temperature of the water begins to drop and may be quite cool for morning use. In such cases, supplemental electric coils can be used to compensate for the heat loss during the nighttime. In many low-latitude nations, solar water heaters are in use. In the southern part of the United States, especially in Florida, the construction of natural gas pipelines brought about the elimination of many solar water heaters, but many remain in the extreme southern part of Florida, which has not yet been reached by the natural gas. Certainly solar water heaters could play a small part in reducing the anticipated demand for electricity predicted in many advanced nations, including the United States.

Solar Heating

Because of the large area of the roof, heating a home with solar energy can be relatively simple if the house is designed and built with this in mind. Here again, water is used to absorb the heat of the sun during the daylight hours. This water is circulated from the rooftop to a storage area of some sort, where the heat may actually be conducted to another liquid or solid material for later retrieval. Problems still exist in producing a system which can store enough heat for extended periods of cloud cover or perhaps extreme cold. Nevertheless, this system could prove useful in reducing the anticipated demand for energy in the future from the traditional energy sources.

Solar Power Plants

Perhaps the most significant potential use of solar energy would be to generate electricity, a power source that is finding increased acceptability by a wider variety of users than ever before. Early sun-following steam generators were built in California and Arizona. One such installation had a collector made of 3,000 individual silvered glass mirrors and was erected in Tempe, Arizona, in 1903. The following year, it was moved to a ranch in Wilcox and presumably was destroyed by a hailstorm. Another interesting application was the construction of the solar generator at the Ostrich Farm in South Pasadena, California, in the early 1900s. Its generated steam was used to pump water and operated for many years.

Today, the two main thrusts are towards the design of small generating systems, which would be located at the points of consumption—the home, the apartment building, the city hall—and the construction of large, central solar power stations, which would feed the electricity generated into the existing grid system.

The latter approach seems to have the most support. Such stations would require large collecting areas, in which the solar heat would be concentrated for transfer to the power plant. In their solar farm proposal, Dr. Aden Meinel, director of the University of Arizona's Optical Sciences Center, and his wife and colleague, Marjorie, suggest that a square of land, seventy-five miles on each side, would provide sufficient area for a power plant having an installed capacity of *one million megawatts.* Of the approximately 5,000 square miles, only 3,000 would actually be covered with collectors. In comparison, that much land is already under lease to provide the strip-mined coal to the infamous Four Corners power plant in New Mexico (see chap. 10), and the Meinel solar power plant would allow for the grazing of sheep among the collectors.

Solar Power in Space

A relatively recent but extremely successful application of solar power has been in space exploration. This was possible due only to the further development of the basic photoelectric cell. The feasibility of direct generation of electricity from solar radiation resulted in the invention of the selenium cell almost 100 years ago. The conversion efficiency of such cells was so low that its use was limited. The most familiar use perhaps, was the selenium light meters used by photographers.

In 1954, it was discovered that properly doped silicon can produce direct current from solar radiation and soon the first such solar batteries were produced. Their conversion efficiency was only about 5 percent but a decade of research and of improved manufacturing techniques has increased this efficiency to as much as 14 percent. Such silicon solar batteries form the basic power sources for the Russian and American space programs, producing sufficient wattage to operate television cameras and radio transmitters, for example.

environmental impact

Negative Impact

A major criticism of solar power plants would be the large area devoted to the collectors, which would be offensive to some. The Meinels answer this by stating that, in the United States, we use more than 500,000 square miles for farming, which produces only 1 percent of our energy needs. We have accepted the idea of agricultural farms and perhaps could eventually accept the idea of solar farms, which would use a small fraction of the area devoted to agriculture today and yet would still allow certain agricultural uses of the land around the collectors. They further state that the 3,125 square miles of land, currently leased by the Bureau of Indian Affairs and the Department of the Interior, would provide enough space for a solar plant capable of generating the one million megawatts expected to be needed by the United States in the year 2000.

Another problem may well be thermal pollution, if water-cooled turbines are used. The threat of thermal pollution would be even greater than with nuclear power plants because of the lower thermal efficiency. Nonetheless, if cooling towers were used, that could help the area restore its local thermal balance, upset by the loss of solar energy to the collectors.

Positive Impact

Solar energy causes no air pollution and needs no mines or wells to supply it with energy. Properly designed, it could even enhance the landscape, certainly more so than many other cultural structures. It would not be depleting any natural resources, with such side effects as waste disposal, mine acid, or subsidence. When one considers the negative aspects of other sources of power, that alone should demand a greater effort by private organizations around the world towards the perfection of the technology which could make solar energy readily available for man's use.

Chapter **14**

Wind Power

Wind has been exploited by man for centuries, most notably to power his ships across lakes, seas, and oceans. The wind represents an important energy source to man, but, because of certain characteristics, its use has diminished during the past century because of competition from other more efficient machinery. The two greatest disadvantages of wind power are the variability of wind speed and wind direction. Apparatus, built to take advantage of the wind, had to be designed so that it could face into the wind for maximum efficiency, or somewhat into the wind to maintain operation. Because of the size and weight of various wind machines, the apparatus could not operate unless a certain minimum wind was available.

Utilization

The most widespread use of wind power has been for sailing. Wind can be somewhat trapped by a variety of materials that can be made into sails. The size, shape, and number of sails determined the amount of energy which could be used to drive a ship. Certainly the magnificent four-masted ships, with their many sails, made efficient use of the wind. But again, the variability of the wind speed and direction placed wind-powered ships at a disadvantage to engine-powered vessels, because the latter had the flexibility to go in any direction without consulting the wind.

The second most widespread use of wind power was in the construction of windmills. These installations could be used to grind grains into flours, saw wood, or pump water. The water-pumping windmills of the Netherlands, used to keep the land, reclaimed from beneath the sea, dry, is what we think of when "windmills" are mentioned. But, even in that little nation, the efficiency and dependability of small, diesel engines which operated the pumps have driven the windmills from the scene. The disappearance of this tourist image has prompted the government to launch a campaign to save some windmills as reminders of a past era.

In the United States, especially out in the Great Plains, the windmill was once as necessary as the house, barn, or silo. Standing like a sentinel, its 8 ft. diameter blades swinging around into the wind to obtain the maximum power, it pumped water day and night, as long as the wind blew, to keep the people, crops, and animals well supplied. Today few working windmills remain in the United States as a reminder of the day when the wind was such an important fact in a person's daily life. In other parts of the world, most notably the Yucatan Peninsula, thousands of functional windmills have thus far withstood the challenge of the diesel engine.

Limitations

The successful development of wind power today depends on the availability of technology which would allow for the development of greater horsepower from a given wind speed. The cost of construction and maintenance, as well as the actual cost of delivered power, would have to compare favorably with alternative sources of power. It now appears that few parts of the world have sufficiently strong winds which blow enough days of the year to justify the development of wind power for that region. Perhaps the Trade Wind Belt, with wind speeds which average 11 to 14 m.p.h. over the Atlantic, offers the greatest potential.

environmental impact

Most will agree that wind power, like solar power, will not adversely affect the environment. Nothing is added to the environment nor taken away. The wind is a renewable, though undependable, resource, and is there for the taking. Certainly the sight of many windmills, especially those with cloth sails, is a delight to behold. However, we can expect little assistance from wind power in meeting the demands for energy while protecting the environment in the years ahead.

Chapter 15

The Consumption of Energy

The consumption of energy in the world increases every year. This is due to the larger, ever-growing world population and to the increased demand for energy for a constantly widening variety of uses, especially in the more advanced nations. Over a period of time, the exact form and type of energy consumed may change as a reflection of such factors as availability, use, technology, and price. In 1969, petroleum was the favorite source of all energy consumed in the world, a position first attained in 1967 (table 15.1). Recent estimates indicate that its share of the market has increased to 54 percent with coal declining to 27 percent.

Petroleum's Popularity

A major reason for this shift to oil has been the passage of legislation, especially in the United States, which is designed to protect the environment. One of the simplest and quickest ways to meet such new regulations

TABLE 15.1

World Energy Consumption by Fuel Type, 1969

(quantities in million tons of coal equivalent)

Fuel Type	Amount	Percentage
Coal	2,563	36
Petroleum	3,015	42
Natural Gas	1,438	20
Hydro & Nuclear Electricity	163	2
World	7,179	100%

SOURCE: *Statistical Papers, Series J, World Energy Supplies*, United Nations, 1959-1969.

was by substituting petroleum products for the tens of millions of tons of coal that otherwise would be burned. The remarkable improvement in air quality in London has in part been due to the elimination of the burning of soft coal within the city. This brought about the increased use of petroleum products, which were easy to obtain and use, to meet the energy requirements of that city.

The United States is the world's leading consumer of petroleum, using almost one-third of the total production (table 15.2). By including oil consumption in western Europe, almost two-thirds of the consumption is accounted for, and the addition of Japan's statistics brings the figure past that mark. Many other nations are turning to oil to meet their burgeoning

TABLE 15.2

World Oil Consumption, 1969

Country or Region	Millions of Tons	Percentage of World Total
United States	735	32
Canada	76	3
Mexico	22	1
Caribbean incl. Venezuela	53	2
Other Western Hemisphere	64	3
Total Western Hemisphere	950	41
Belgium-Luxembourg	28	1
Netherlands	36	2
France	88	4
West Germany	130	6
Italy	86	4
United Kingdom	106	5
Sweden	54	2
Other Western Europe	89	4
Total Western Europe	617	28
Middle East	50	2
Africa	43	2
South Asia	30	1
Southeast Asia	56	2
Japan	178	8
Australasia	30	1
U.S.S.R., Eastern Europe, and China	342	15
Total Eastern Hemisphere	1,346	59
World	2,296	100%

SOURCE: Reprinted by permission, © *Encyclopaedia Britannica*, 1972.

energy demands, and this is helping to push the total amount of oil consumed even higher. Japan illustrates this situation, for her coal and hydroelectric resources have proved far inadequate to meet the energy demands that have accompanied her phenomenal economic development of the past two decades. Japan therefore turned to oil and to the design and construction of VLCC's to transport the prodigious amounts of oil needed from the Persian Gulf region.

refining capacity

For most uses, crude petroleum must be refined into by-products. This situation demands that many refineries be located throughout the world. Presently, the United States contains 25 percent of the world's refining capacity (table 15.3). Since she consumes 32 percent of the oil, about 7 percent of the by-products must be imported from foreign refineries. Table 15.3 indicates the rather broad dispersal of refinery capacity throughout the world, in both oil exporting and oil importing nations.

In the refining process, there is no weight loss. Thus, for every barrel of oil refined, a barrel of by-products is produced. This allows refineries a freedom of location not enjoyed by most other mineral processing installations. Refineries are resource oriented, break-of-bulk oriented, or market oriented. In the post-World War II period, the emphasis has been on market oriented refineries which could produce the proper ratio of refined products for the nearby markets, while importing oil from areas throughout the world. Sometimes the market oriented and break-of-bulk oriented refineries coincide, as in the refining complexes of Rotterdam, Netherlands. The by-products are produced for the local market in that country, but, because its demands are relatively small, the refining capacity was enlarged to meet the market needs of a vast hinterland served by pipelines, barges, and railroads. So both locational factors are represented there.

More recently there has been an emphasis, especially in the more advanced nations such as the United States, to discourage the construction of refineries because of their negative impact on the environment. At the present time, there is just one refinery under construction in the United States. It is being built by Mobil in Joliet, Illinois. In the meantime, the demand for refined products continues to increase, and the United States is becoming more and more dependent upon foreign refineries. To the American environmentalists, this is perfectly alright, but it does place the United States in a more vulnerable position by its inability to refine all the oil it can obtain.

Foreign nations are now beginning to resist the use of their land for the construction of refineries which will harm their environment while serving markets in the United States and other advanced nations. Thus,

TABLE 15.3

World Oil Refining Capacity in 1969

Country/Area	Total tons (in 000,000s)	Percentage of Total
North America		
United States	647	25.6
Canada	66	2.6
Mexico	30	1.1
Total North America	743	29.3
Caribbean (incl. Venezuela)	186	6.9
Other Western Hemisphere	65	2.6
Total Western Hemisphere	994	38.8
Western Europe		
Belgium	35	1.4
France	120	4.7
West Germany	127	4.9
Netherlands	63	2.5
Italy	146	5.7
United Kingdom	119	4.6
Others	120	4.7
Total Western Europe	730	28.5
Middle East	121	4.7
Africa	40	1.5
South Asia	34	1.3
Southeast Asia	59	2.3
Japan	157	6.2
Australasia	33	1.3
U.S.S.R., Eastern Europe and China	395	15.4
Total Eastern Hemisphere	1,569	61.2
World Total	2,563	100.0%

SOURCE: Reprinted by permission, © *Encyclopaedia Britannica*, 1971.

heated discussions are sure to crop up around the nation as proposals are presented for the construction of superports, tank farms, and new refineries.

Gas is Best

This familiar slogan has long advocated the superior qualities of natural gas. And, indeed, natural gas consumption has greatly increased, especially in recent years. For example, during the period 1959-69, natural gas increased its share of the world energy market from 14 to 20 percent. This can be explained by its increased availability, minimal effect on the environment, and in many cases, lower cost. In many nations, the demand greatly exceeds the supply, and the world is just now witnessing the construction of facilities which will move liquified natural gas around the world to waiting markets, just as the oil industry has been doing for years.

In many parts of the United States, the demand for natural gas is much greater than the available supply. As a result, installation of equipment using natural gas has been forbidden, quota systems established, or, in some cases classes of consumers have been established to give certain economic activities priority over others. In the State of Ohio, food processing firms are usually able to convert existing equipment over to natural gas because of the desirability of keeping the plant as clean as possible. During the winter months, many large consumers have their natural gas supplies reduced or suspended, which often results in some having to close down their operations until the shortage of fuel supply eases. This happened to the Lordstown assembly plant belonging to General Motors Corporation. Their solution was quite unusual. An enterprising geologist offered to locate natural gas for them, and upon acceptance of his offer, he indicated a spot in the parking lot where a well should be dug. And natural gas was discovered which can meet the plant's energy need. Many other large consumers around the country would like to meet their demand for natural gas that easily!

The Popularity of Electricity

During the decade of the 1970s, the consumption of electricity is expected to equal the total amount consumed from the beginning of the electric power industry in the 1880s through the end of 1969 (table 15.4). This illustrates the increased preference for electricity in industrial, commercial, and residential applications which once were met directly by such energy sources as coal, manufactured gas, or water power.

While the world's preference for electricity, as the final form in which the energy will be consumed, continues to increase, the primary energy mix used to produce this electricity has changed over time and is expected to show distinct shifts by 1990. For example, in 1970, 46 percent of the electric power generated in the United States was produced by coal, 25 percent by natural gas, 12 percent by petroleum products, 15 percent by waterpower, and almost 2 percent by nuclear energy. In 1990, it is predicted that almost half of the electricity produced will be generated by nuclear power, coal will produce 21 percent, natural gas 14 percent, petroleum products 9 percent, and waterpower 7 percent. With the existing controversy over the actual safety of the largest nuclear power plants operating and under construction today, this prediction may prove to be far

TABLE 15.4

Production of Electric Energy by Central Stations in the United States, 1902-70

(millions of kilowatt-hours)

Year	Total	Hydro	Percentage	Fuel	Percentage
1902	5,969	2,166	36	2,803	64
1912	24,752	7,387	30	17,365	70
1920	56,559	20,311	36	36,248	64
1925	84,666	26,112	31	58,554	69
1930	114,637	34,874	30	79,763	70
1935	118,935	42,253	36	76,682	64
1940	179,907	51,659	31	124,941	69
1945	271,255	84,747	33	181,708	67
1950	388,674	100,884	28	281,000	72
1955	629,010	116,236	20	502,388	80
1960	841,616	149,123	18	692,494	82
1965	1,157,583	196,984	17	960,599	83
1970	1,639,771	247,456	15	1,392,315	85

SOURCE: *Statistical Abstract of the United States, 1972; Historical Statistics of the United States, Colonial Times to 1957*, (Washington, D.C.: U.S. Department of Commerce).

TABLE 15.5

Projected Installed Capacity of Nuclear Power Plants in 1976

Country	Number of reactors	Total power capacity (megawatts)
United States	107	76,500
United Kingdom	42	13,000
West Germany	25	10,000
Japan	19	9,500
U.S.S.R.	29	8,800
Sweden	10	5,400
Canada	9	4,000
France	11	3,000

SOURCE: *The Americana Annual*, 1971.

from accurate (table 15.5). Nuclear moratoriums may prevent a wide application of nuclear generation, and new energy sources, such as solar and geothermal installations, may contribute in a significant way. So we have an indication of the many variables which exist in forecasting energy demands in the future, not just because of the changing mix of present and potential energy sources, but also because of the large number of countries with fluctuating population and economic growth rates, as well as environmental restrictions. For example, note the changes that have taken place in the United States energy picture between 1940 and 1969 (table 15.6).

TABLE 15.6

United States Mineral Fuel Resources and Electricity Production and Consumption; 1940, 1955, 1969

(in trillions of British thermal units)

Year	Total	COAL		Crude petroleum	Natural gas	Electricity	PERCENTAGES				
							COAL				
		Anthracite	Bituminous and Lignite				Anthracite	Bituminous and Lignite	Crude petroleum	Natural gas	Electricity
PRODUCTION											
1940	25,088	1,308	12,072	7,849	2,979	(NA)	5.2	48.1	31.3	11.9	(NA)
1955	38,900	665	12,174	14,410	10,204	(NA)	1.7	31.3	37.1	26.2	(NA)
1969	59,587	266	14,685	18,922	22,858	2,856	0.4	24.6	31.8	38.4	4.8
CONSUMPTION											
1940	23,908	1,245	11,290	7,487	2,969	(NA)	5.2	47.2	31.4	12.4	(NA)
1955	39,956	599	11,104	16,328	10,428	(NA)	1.5	27.8	40.8	26.1	(NA)
1969	65,773	245	13,293	28,419	21,040	2,776	0.4	20.2	43.2	32.0	4.2

SOURCE: *Statistical Abstract of the United States,* (Washington, D.C.: U.S. Department of Commerce, 1971).

NA Not available

Table 15.7

World Consumption of Total Industrial and Public Electricity, 1970

(quantities in thousand million kilowatt hours, and in kilowatt hours per capita)

Country	Total Consumption	Consumption Per Capita
United States	1,640.264	8,009
U.S.S.R.	735.626	3,030
Japan	350.590	3,391
West Germany	250.363	4,067
United Kingdom	249.222	4,464
Canada	201.300	9,404
France	140.209	2,759
Italy	121.388	2,262
China	93.554	121
East Germany	68.052	3,943
Sweden	64.703	8,042
Norway	56.770	14,635
Australia	56.153	4,474
Czechoslovakia	49.368	3,412
Netherlands	40.518	3,110
Belgium-Luxembourg	34.591	3,454
Switzerland	28.413	4,509
Finland	23.078	4,915
New Zealand	13.815	4,916
Other	694.129	—
World Total	4,912.106	1,361

Source: *World Energy Studies*, 1961-1970, (New York: United Nations, Statistical Office, 1972).

per capita consumption

The statistics in table 15.7 indicate the leading producers of electricity in the world. As might be expected, the United States produces more than twice as much electricity as any other nation, and accounts for about one-third of the world's total production. But, surprisingly, it ranks third in per capita consumption of electricity. Norway's prominent position in first place is due to the metallurgical industries located along the northern coast. Because of their large demands for electricity, and Norway's relatively small total population, the per capita consumption for this Scandinavian country is so very large. But in a real sense, the per capita consumption by private individuals is not very great.

regional consumption

World energy consumption is dominated by two regions—North America and the U.S.S.R. and its satellite nations—with a combined total of 65 percent of the energy consumed in 1969 (table 15.7). By adding the 19 percent consumed by western Europe, these three regions account for 84 percent. Thus, less than one-third of the world's population used 84 percent of all energy consumed in 1969.

interregional trade

North America, western Europe, and the Far East were net importers of energy in 1969. That means that Latin America, western Asia, and the U.S.S.R. and its satellite nations were the net exporters of energy, mainly due to the surplus production of petroleum. However, smaller regions and individual nations may depart from such broad generalizations. For example, the United States, although a net importer of energy, exports considerable quantities of coal to other continents, while California exports petroleum products to many western states but imports natural gas from fields in Canada, Texas, and New Mexico, and obtains hydroelectricity from the Columbia River Basin.

Meeting Future Demands for Energy

What can we do about future demands for energy? It appears certain that the advanced nations cannot continue to consume such a majority of the total energy each year, if the level of living is to be increased in the underdeveloped nations around the world. Certainly, more energy will be needed by the world, but a larger percentage of this will have to be consumed by the developing and underdeveloped nations. Much evidence

TABLE 15.8

World Energy Production and Consumption, 1970

(quantities in million tons of coal equivalent)

	Production of Total Energy	Consumption of Total Energy
World	7699	7521
Africa	496	119
North America	2486	2719
Latin America	470	281
Far East	259	578
Middle East	1035	85
Centrally-Planned Asia	471	484
Centrally-Planned Europe	1793	1687
Western Europe	612	1487
Oceania	78	85

SOURCE: *World Energy Studies*, 1961-1970, (New York: United Nations, Statistical Office, 1972).

exists which supports the notion that the higher the consumption of energy per capita, the more advanced is the economy of the nation. Therefore, energy in one form or another will be applied to improve opportunities for progress around the world.

One avenue of attack is simply to find more of the traditional energy sources by digging more coal and finding more oil and natural gas. Although tremendous amounts of these fuels lie yet to be discovered, the scope of needed supplies is challenging indeed. The large reserves of oil available in Alaska would last the United States only three years, based on the 1970 rate of petroleum consumption.

A second avenue is to develop nuclear energy. This includes locating new sources of uranium and investing in larger nuclear power plants to reduce costs of fission generation, along with the perfection of more advanced technologies such as fusion and the breeder reactor. This is the direction which the United States has chosen to take, and towards which it is well committed. Today, however, despite this commitment, opposition seems to be mounting, challenging both the wisdom of the choice, and the safety and feasibility of nuclear power plants.

A third avenue which should be thought of as a supplementary course is the development of new energy sources such as solar and geothermal power. Certainly, calculations suggest that they could supply a significant portion of the electricity needed, within two to three decades. Given technological breakthroughs, wind and tidal power could also contribute energy to some regions.

Energy Conservation

An avenue of attack that can have an immediate effect on the energy situation is along the line of conserving energy. The United States in particular has developed a way of life based upon the abundance and low cost of energy, and the present shortages and increasing costs of energy are alerting people to the need to do things differently.

architecture

As in any era, a particular architectural style comes into vogue, and such style sells itself. In many of today's buildings, the style has been greatly influenced by a desire for minimum maintenance and maximum protection against vandalism. Thus, many schools and other public buildings are relying on more electric lighting and air conditioning to create a pleasant indoor environment while eliminating the windows of yesterday which let in fresh air and light. This trend demands more energy.

In private homes and many commercial buildings, the emphasis has been to bring the outside in, or to let the public see the inside of the establishment as a form of advertising. This requires the use of much glass, and glass is a notoriously poor insulator. In winter, tremendous amounts of energy are lost to the atmosphere from houses and other buildings featuring sliding glass doors, huge glass windows, and insufficient insulation.

It is also a known fact that winter heating and summer cooling costs can be reduced by siting a house with proper consideration for the prevailing winter winds or the direction of the rays of the summer sun. The challenge here is getting the builder to consider this benefit to the future owner when he subdivides his expensive land and positions the houses.

air conditioning

A major cause of the electrical brownouts and blackouts during the summer months is the widespread popularity of air conditioning. Its use is to be expected in the southern parts of the United States, but, with the increased affluence and intense use of urban land for residential and commercial activities, it has become increasingly common throughout the United States. Whereas once electric power companies were stuck with excess capacity during the summer months, their advertising campaigns, along with those of the manufacturers of air conditioning equipment, encouraged its acceptance to such a degree that a lack of sufficient capacity during the summer months is today a chronic problem. Now that people are accustomed to a man-modified climate, it is difficult to get them to change back to the less comfortable conditions of yesterday. Controlled indoor temperature has been proven to improve the performance of both the worker and the student. It allows for the reduction of dust in the air and easier housekeeping, as well as the ability to control the humidity in the air. Thus, many will argue that new energy sources must be developed to support this better way of life, rather than to ask man to step backwards.

If one is used to air conditioning in his home and place of work, should not he expect air-conditioned transportation? Indeed that was the logical extension of this change in living and working conditions and in the post-World War II period, the American automobile has largely become air-conditioned, especially in the hotter climates. This drain on the engine to operate the air conditioning equipment has demanded larger engines or higher rpm's, both of which reduce the mileage per gallon of fuel. Small, economical engines which may give adequate performance for their owners become sluggish, underpowered, and even dangerous when fitted with an air conditioner. Whether in the home, in the car, or at work, air conditioning, with its increased comfort and demand for energy, has become an accepted aspect of everyday life in the United States.

relative transportation costs

In moving goods and people, the world has emphasized convenience, speed, and flexibility. Thus, in many regions truck transportation has replaced rail transport, even though trucks are only one-fourth as efficient in terms of energy use, and produce more air pollution. Airplanes have encouraged the dispersal of economic activities throughout this country and the world, becoming a most significant transporter of people, mail, and low-value, high-bulk goods. Its advantage is speed, but it provides this at 1/60th the efficiency of trains.

The continued popularity of water transport (barges and ships), and the widespread acceptance of pipeline transport during the past three decades, reflect their efficient use of energy. With these comparisons in mind, it is easy to see how transport costs could be reduced by using railroads, pipelines, and water transport to a greater extent, especially for freight, while public transport (buses, commuter trains, subways) could reduce the energy demand now represented by large-engined, air-conditioned automobiles.

At the present time, serious gasoline shortages face the United States. Coupled with the higher cost of the petroleum, the smaller car is becoming more desirable.

For example, a 5,000-pound car burns about twice the fuel needed to propel a car weighing 2,000 pounds. With the different sizes of cars now operating on the highways, the reduction of speed limits would also cut fuel consumption. Cars traveling seventy-five to eighty miles per hour use perhaps 20 percent more fuel than is needed to go fifty miles per hour. The federal government has emergency plans to reduce speed limits on federal highways if the fuel supply situation dictates the need for such measures.

exterior lighting

It has been proven that, in most cases, better illuminated streets, parking lots, alleys, and yards around homes reduce the crime rate. In this country, we have taken the easy way out to try to solve this problem—using cheap electricity to discourage crime rather than by attacking the root of the problem. Faster travel speeds on streets and highways have demanded levels of illumination at intersections and other busy sections of the nation's roadways, to enable the drivers to see far enough ahead in order to properly control their vehicles. Again this means more energy consumption.

advertising with light

The use of electric lights to advertise a place of business has long been an accepted practice in most parts of the world. In some cities, the fabulous variety, design, and number of illuminated signs might even be considered excessive, as in Las Vegas, Nevada, or the Ginza district in Tokyo. In many cities, it is now considered fashionable to leave your store or office fully illuminated so that those who pass by can look in and get a positive feeling toward your enterprise. Everywhere one looks—used and new car lots, outdoor billboards, restaurants, drive-ins, etc.—one sees a greater demand for electricity to outdo someone else's sign in order to attract more business. Hopefully this present energy shortage will encourage the adoption of smaller signs needing less energy. Not only will it save businesses money and reduce the demand for energy, but it will help clean up the landscape and cityscape which has become so cluttered with signs.

energy in the home

As our way of life has changed since the Second World War, a noticeable increase in energy consumption has taken place in homes and apartments. As mentioned earlier, air conditioning is the prime example of this increased energy usage. But then there are washers and dryers, color TV's, self-defrosting refrigerators and an endless list of other appliances. It has been estimated that the pilot light in a gas stove consumes about 10 percent of the gas used by that stove during the year. Today we heat all the rooms in the home, whether in use or not. Likewise we keep them well lit at night and cooled during the summer months, and, with the increased size of living space per capita, the consumption of energy has kept on growing.

Every one of us can cite different examples of energy use that affects his or her life in a positive or negative way: night baseball for professionals or little leaguers; night golf at driving ranges or golf courses; more comfortable homes, schools, and places of work due to higher light levels, air conditioning, and ample space heating; and quick and easy transportation via car or jet airplane. Could you give up, or restrict your use of any of these? Obviously, each of us has his own most desired energy applications, as well as his own ideas as to which types of energy consumption should be reduced.

Major Consumers of Energy

In the period of 1950-70, total energy consumption in the United States doubled. In the same period, population increased by only 40 percent, leaving increased affluence to account for the other 60 percent of the energy consumption increase. The consumption of electricity has doubled every ten years in this country. Now how is this energy being consumed in one form or another in the United States?

The major consumer of energy is transportation, accounting for 25 percent of the total American consumption. Obviously, this represents a prime target for increased efficiency of use, whether it be by train, mass transit in urban areas, smaller, more economical cars, or decreased inter-urban truck transportation. The second major use of energy is for space heating, larger homes, offices, manufacturing plants, and commercial enterprises create a need for more heat, especially with our preference for warmer temperatures. Process steam and direct heat together represent 27 percent of the total energy budget. This area involving the production of so many products and by-products also attracts attention as one in which economies can be realized. Other uses are relatively small but that does not mean that economies cannot be effected, or that different energy sources could not be utilized.

The situation suggests the need for a reevaluation of our energy resources and our energy uses. Certainly, it is most important to find ways to reduce the rate of consumption increase. Modifications in our life styles can also affect economies. But the basic problem still remains that of man, his thoughts, and his actions. If we can get him to change his thinking, to stop seeking prestige, wealth, and satisfaction through the consumption and possession of material goods, we can improve the quality of life while making lesser demands on the world's energy resources. This should be the combined goal of each one of us now.

Index